Electromagnetic Shielding and Corrosion Protection for Aerospace Vehicles

Jan W. Gooch
John K. Daher

Electromagnetic Shielding and Corrosion Protection for Aerospace Vehicles

 Springer

Jan W. Gooch
Georgia Institute of Technology, Atlanta
2020 Howell Mill Road
Atlanta, GA 30318
USA
Email: jangooch@yahoo.com

John K. Daher
Georgia Institute of Technology, Atlanta
3028 Gant Quarters Circle
Marietta, GA 30068
USA
Email: john.daher@gtri.gatech.edu

Library of Congress Control Number: 2006938807

ISBN 978-1-4419-2358-5   e-ISBN 978-0-387-46096-3

Printed on acid-free paper.

9 8 7 6 5 4 3 2 1

springer.com

Contents

Preface...ix

Acknowledgment..xi

1. Electromagnetic Shielding Effectiveness and Corrosion Prevention...............1
   1.1 Electromagnetic Environment...1
   1.2 Importance of Electromagnetic Shielding and Corrosion
        Prevention for Aircraft 1

2. Fundamentals of Corrosion 5
   2.1 Corrosion 5
   2.2 Conditions for Corrosion 6
        2.2.1 Electrochemical Aspects 6
        2.2.2 Environmental Aspects 8
        2.2.3 Metallurgical Aspects 10
   2.3 Types of Corrosion 11
        2.3.1 Uniform Attack 11
        2.3.2 Galvanic or Two-Metal Corrosion 12
        2.3.3 Crevice Corrosion 12
        2.3.4 Pitting 13
        2.3.5 Intergranular Corrosion 14
        2.3.6 Selective Leaching 14
        2.3.7. Erosion Corrosion 14
        2.3.8 Stress Corrosion 14

3. Fundamentals of Electromagnetic Shielding 17
   3.1 Definition of Shielding Effectiveness 17
   3.2 Factors that Determine Shielding Effectiveness 17
   3.3 Electromagnetic Shielding Theory..18
        3.3.1 Absorption Loss 18

3.3.2   Reflection Loss...20
3.3.3   Multiple Reflection Correction Term..........................21
3.3.4   Effect of Near-Field Sources on Shielding Effectiveness.............21
3.4   Practical Electromagnetic Shielding 22
3.5   Role of Conductive Sealants in Electromagnetic Shielding 24

4.   **Investigation of the Relationship Between DC Resistance and
     Shielding Effectiveness** **25**
4.1   Determining a Relationship Between DC Resistance and Shielding
      Effectiveness 25
4.2   Stainless Steel Test Joint for DC Resistance and Shielding
      Effectiveness Measurements..25
4.3   DC Resistance Measurements...28
4.4   Shielding Effectiveness Measurements....................................29
4.5   Measurement Results on the Relationship Between DC Resistance
      and Shielding Effectiveness 30
4.6   Effect of Plate Separation on Shielding Effectiveness................33

5.   **Identification and Evaluation of Optimum Conductive Sealant
     Materials** **35**
5.1   Identification of Materials...35
5.2   Evaluation of Conductive Sealants in Salt Spray Environment per
      ASTM B117 38
      5.2.1   Aluminum Test Joints 38
      5.2.2   Measurement Results of DC Resistance Versus Salt Spray
              Exposure..42
      5.2.3   Measurement Results of Shielding Effectiveness Versus
              Salt Spray Exposure 43
5.3   Summary of Optimum Conductive Sealant Materials.................46
5.4   Relationship Between DC Resistance, Shielding Effectiveness, and
      Transfer Impedance...46
      5.4.1   Evaluation of Laboratory Sealant Test Samples.............46

6.   **Field Test Evaluations on E-3A Aircraft**....................................**57**
6.1   Introduction 57
      6.1.2   Visits to Air Force Installations 58
6.2   Field Test Description 63
      6.2.1   Location of Test Sites..63
      6.2.2   Field Test Methods...63
      6.2.3 E-3A Aircraft Field Test Results...............................72
6.3   Conclusions 73
6.4   Recommendations 74

7 **Assessment of the Validity of the MIL-B-50878 Class R Bonding Requirements**...**75**
  7.1 Background 75
  7.2 Basis/Rational Behind the 2.5-Milliohm Bonding Requirements 76
  7.3 Summary and Conclusions on Validity of 2.5 Milliohm Bonding Requirements...78
  7.4 Proposed Changes to Standards, Specification and Handbooks.................78
      7.4.1 Modifications to MIL-B-5087B..80
      7.4.2 Modifications to MIL-STD-188-124A 82
      7.4.3 Modifications to MIL-STD-1541..85
      7.4.4 Modifications to MIL-STD-1542..86
      7.4.5 Modifications to MIL-STD-454K..87
      7.4.6 Modifications to MIL-HDBK-253..88
      7.4.7 Modifications to AFSC Design Handbook 1-4..............................91
      7.4.8 Modifications to NATO STANAG No. 3659......................................97
  7.5 Recommendations for Corrosion Prevention Materials/Processes for Use on Existing (and Future) Aircraft and Weapon Systems 99
  7.6 Recommendations for Using Corrosion Prevention Materials for retrofitting Existing Nuclear Harden Aircraft and Weapons Systems...99
      7.6.1 Retrofitting Aircraft with Conductive Sealant................................99
      7.6.2. Annodized and Chemical Conversion Type Treated Aluminum Surfaces...99
  7.7 Resealable Joints Using Liquid Applied Conductive Sealants100
      7.7.1 Resealable Joints in Salt Spray Chamber......................................100
      7.7.2 Opening Resealable Joints 101
      7.7.3 Comments...101
  7.8 Shielding Effectiveness of Non-Conductive Corrosion Inhibitive Sealant...101
  7.9 Aerospace Material Specification 102

8. **EMI Gaskets**...**103**
  8.1 EMI Gasket Applications and Selection Criteria....................................103
  8.2 Stamped and Formed Metal Gaskets..103
  8.3 Wire Mesh Gaskets 104
  8.4 Conductively-Loaded Elastomeric Gaskets...105
  8.5 Conductive Fabric-Clad Foam Gaskets 105
  8.6 Conductive Gel Gaskets..106

A. Shielding Effectiveness Data for Test Joints with Varying
   Resistances 107

B. Shielding Effectiveness Versus Plate Separation Data 117

References 123

Index 125

Preface

The new generation of civil aircraft depends heavily on electronic systems to implement safety-critical functions. Because these aircraft may be exposed to high intensity radiated fields (HIRF) created by radio frequency (RF) emitters based on the ground, in the air, and at sea, civil aviation authorities have become increasingly concerned about the potential for electromagnetic interference (EMI) to these critical electronic systems. With the increasing density of the electromagnetic environment, civilian aircraft authorities and standards committees have developed HIRF requirements that are being applied to civil aircraft certification programs.

Military aircraft and weapon systems must operate compatibly within an electromagnetic environment that can be even more severe than the civil HIRF environment. For example, the electric fields from a high altitude, nuclear electromagnetic pulse (EMP) can reach a peak amplitude of 50,000 V/m. Emerging high power microwave (HPM) threats can generate power densities in excess of 100 W/cm^2 at tactical ranges. And aircraft that must take off from and land on naval ships can be exposed to a dense and highly intense electromagnetic environment with electric field strengths exceeding 1000 V/m.

Military and commercial aircraft are assembled with many fasteners such as screws and bolts to secure airframes, skin and other bonded joints. The joints must be conductive or shielding effectiveness will not be optimized. Conductive and corrosion resistant materials such as sealants and gaskets are required to maintain conductivity along the entire structure of the aircraft under flight and ground environmental conditions including moisture, rain, and pollution from the atmosphere. The conductivity of corrosion resistant sealant materials must be monitored under accelerated weathering conditions to determine the conductivity over an extrapolated period of time to simulate an acceptable life of the sealant between scheduled maintenance periods. Comprehensive studies that provided understanding of shielding effectiveness s as related to conductivity, and the relationship of material chemistry to conductivity and corrosion were performed to provide fundamental understanding of these phenomena, and finally, optimized shielding effectiveness for aircraft.

Significant progress was made in the identification and, ultimately, optimization of corrosion prevention materials and/or processes. These materials are capable of protecting metal surfaces from air/moisture corrosion over a specified period of time. In addition, electrically conductive corrosion prevention materials that are capable of maintaining EMI/EMP protection of aircraft and weapon systems were identified. With these capabilities, existing aircraft and weapon systems can now be protected from further corrosion, and existing nuclear-hardened weapon systems can be retrofitted with corrosion prevention. From these efforts, American Materials Standard, AMS 3262, was prepared for the user of conductive silicone rubber-based sealant.

Acknowledgment

The authors wish to acknowledge David Ellicks and Eugene Bishop of the Warner Robins Air Logistics Center, U. S. Air Force, for their significant contribution to the research and development projects that provided information for this book, and for their continued interest in electromagnetic shielding and corrosion prevention to provide a safer operating environment and maximum operating lifetime for aircraft. In addition, the authors wish to express their gratitude to the Society of Automotive Engineers (G9 Committee) for their cooperation during the preparation, acceptance and publication of American Materials Standard, AMS 3262, SEALING COMPOUND, SILICONE RUBBER, TWO-PART, ELECTRICALLY CONDUCTIVE AND CORROSION RESISTANT, FOR USE FROM –67 to 500 °F (–55 to 260 °C). AMS 3262 was a product of the information generated from the projects discussed in this book.

1

Electromagnetic Shielding Effectiveness and Corrosion Prevention

1.1 Electromagnetic Environment

The new generation of civil aircraft depends heavily on electronic systems to implement safety-critical functions. Because these aircraft may be exposed to high intensity radiated fields (HIRF) created by radio frequency (RF) emitters based on the ground, in the air, and at sea, civil aviation authorities have become increasingly concerned about the potential for electromagnetic interference (EMI) to these critical electronic systems. With the increasing density of the electromagnetic environment, civilian aircraft authorities and standards committees have developed HIRF requirements that are being applied to civil aircraft certification programs.

Military aircraft and weapon systems must operate compatibly within an electromagnetic environment that can be even more severe than the civil HIRF environment. For example, the electric fields from a high altitude, nuclear electromagnetic pulse (EMP) can reach a peak amplitude of 50,000 V/m. Emerging high power microwave (HPM) threats can generate power densities in excess of 100 W/cm^2 at tactical ranges. And aircraft that must take off from and land on naval ships can be exposed to a dense and highly intense electromagnetic environment with electric field strengths exceeding 1000 V/m.

1.2 Importance of Electromagnetic Shielding and Corrosion Prevention for Aircraft

To provide a layer of isolation between the external electromagnetic fields and the internal electronic systems, metal panels surrounding the electronic systems are fastened together with low impedance bonds to form a shield. Examples of electromagnetic shields include the aircraft or missile skin, shields around avionics bays, and shields around individual electronic boxes. Under present service conditions, aircraft and weapon systems can experience corrosion between these metal surfaces. This problem is two-fold. First, corrosion between metal surfaces creates structural weaknesses that undermine the effectiveness of the structure. Secondly, the corrosive process produces non-conductive products, which degrade the electromagnetic shielding and/or nuclear-hardening capability of the structure.

A primary requirement for effective electrical bonding is that a low impedance path be established between the two panels to be joined. Procurement specifications commonly use electrical resistance rather than impedance as the specified parameter. The limiting value of resistance at a particular junction is a function of the bonding requirements and the amount of current expected to pass though the path. For example, where the bond serves only to prevent static charge buildup, relative high bond resistances, i.e., on the order of 1 Ω, may be acceptable. On the other hand, bonds for shock hazard protection are commonly limited to 0.1-Ω resistance. Where RF interference may result from a poor bond between equipment and its mounting surface, the bond resistance is limited by MIL-B-5087 and MIL-STD-181 to 2.5 mΩ. Moreover, to be effective over the lifetime of the system, the impedance of this path must remain low with use and with time.

The imposition of low values of bond resistance helps ensure that impurities are removed from the mating surfaces and that sufficient surface contact area is provided to establish a reliable path for currents to flow while limiting voltage differentials across the bond junction. It is these voltage differentials that lead to unwanted energy coupling into protected volumes. However, it is widely recognized that a low dc resistance is not a reliable indicator of the performance of the bond at high frequencies such as those in the upper portions of the EMP spectrum, in the VHF/UHF communications band, and in the radar frequency bands. Inherent bonding impedance, which includes inductive and capacitive reactances, will determine the total impedance of the bond. An additional factor in bond performance is the bond aperture dimensions relative to the wavelength of incident energy. As an example, a spot welded bond can easily meet the 2.5-mΩ requirement and, yet, if the welds are widely spaced, appreciable electromagnetic energy can reach the interior of the bonded region. When exposed to severe electromagnetic environments, the coupled energy can produce upset or damage of aircraft avionics equipment. It is for this reason that most EMP control documents do not specify a resistance performance criterion for bonds—they, instead, specify the method of construction (typically continuously welded).

Historically, the MIL-C-5541 Class 3 chemical conversion coating procedure used on aluminum surfaces has been the only protective finish that meets the existing EMP/EMI requirements. However, this method allows moisture to be trapped between two metal surfaces, which causes corrosion of the surfaces. The products of corrosion are non-conductive materials which increase the electrical resistance of the bond or joint between surfaces, resulting in an RF impedance high enough to degrade the shield effectiveness of shielded aircraft structures and to destroy the nuclear hardness of many aircraft and weapon systems.

From the above description of the problem, it is apparent that corrosion protection materials and/or processes are required which are compatible with EMI and nuclear EMP hardened aircraft and weapon systems. Furthermore, with the evident difficulty of achieving the 2.5 mΩ dc resistance while maintaining a corrosion-free bond for an extended period, there clearly is a definite need to seriously examine the relative need of such a low resistance in order to produce a HIRF certified or an EMI/EMP-hardened system. An iterative approach is, therefore, taken which is structured toward establishing a carefully planned and documented technology base that demonstrates the relationship between bond dc resistance, corrosion protection and the electromagnetic protection offered by the bond.

The above efforts require identification and, ultimately, optimization of corrosion prevention materials and/or processes. These materials must necessarily be capable of protecting metal surfaces from air/moisture corrosion over a specified period of time. In addition, electrically conductive corrosion prevention materials that are capable of maintaining EMI/EMP protection of aircraft and weapon systems need to be identified. With these capabilities, existing aircraft and weapon systems can be protected from further corrosion, and existing nuclear-hardened weapon systems can be retrofitted with corrosion prevention materials.

2

Fundamentals of Corrosion

2.1 Corrosion

Corrosion can be defined in several ways: (1) destruction or deterioration of the material because of reaction with its environment; (2) destruction of materials by means other than straight mechanical; (3) and extractive metallurgy in reverse [1]. Definitions 1 and 2 are preferred for purposes of this book because the corrosion of aluminum used in the manufacture of aircraft is of primary importance. Another definition [2] is: Corrosion is the result of an electrochemical process involving an anodic reaction, the metal goes into solution as an ion, and a cathodic reaction takes place where the electrons released by the anodic reaction are discharged to maintain electrical neutrality by reaction with ions in solution. Corrosion resistance or chemical resistance depends on many factors, and its complete and comprehensive study requires knowledge of several fields of sciences including chemistry, metallurgy and biology when microorganisms deteriorate the surfaces of metals.

Aluminum is a reactive lightweight metal (low density compared to iron), but it develops an aluminum oxide coating or film that protects it from corrosion in many environments. The aluminum oxide film (Al_2O_3) is stable in neutral and many acid solutions but is attacked by alkalis. The active aluminum surfaces of aircraft parts are often protected with chromate conversion coatings. In addition to corrosion resistance, other properties include colorless and nontoxic corrosion products, appearance, electrical and thermal conductivity, reflectivity, and good strength to weight ratio. Pure aluminum is soft and weak, but is be alloyed and heat-treated to a broad range of mechanical properties that comprise aluminum components for aircraft.

Like all technologies, corrosion science has its own jargon. The material or agent attacking a metal or other substrate is generally referred to as the corrosive and the material under chemical attack is the corrodent and the process is referred to as corrosion. For example, a corrodent is said to be corroded by a corrosive. The National Association of Corrosion Engineers [3] is an excellent source of additional published technical information and conferences the subject of corrosion.

2.2 Conditions for Corrosion

There exists important conditions in the natural environment and service in an actual aircraft that favor corrosion of most metals including:

1. Electrochemical aspects
   (a) Electrochemical reactions
   (b) Polarization
   (c) Passivity
2. Environmental aspects
   (a) Effect of oxygen and oxidizers
   (b) Effects of velocity
   (c) Effects of temperature
   (d) Effects of corrosive concentration
   (e) Effect of galvanic coupling
   (f) Corrosion from biological organisms
3. Metallurgical aspects

   A useful summary of each aspect is discussed below to assist the reader in understanding the fundamentals of corrosion that play an important part in selecting materials of construction, including aircraft and materials that are discussed. References are given for more extensive reading on each subject.

2.2.1 Electrochemical Aspects

1. Electrochemical Reactions. The electrochemical nature of corrosion can be illustrated by the attack of a reactive metal such as zinc by hydrochloric acid. When zinc is placed in hydrochloric acid, a vigorous reaction occurs; hydrogen gas is evolved and the zinc dissolves, forming a solution of zinc chloride. The reaction is:

$$Zn + 2HCl \rightarrow ZnCl_2 + H \qquad 2.1$$

   Noting that the chloride ion is not involved in the reaction, this equation can be rewritten as:

$$Zn + 2H^+ \rightarrow Zn^{+2} + H_2 \qquad 2.2$$

   Therefore, zinc reacts with the hydrogen ions of the acid solution to form zinc ions and hydrogen gas. Examining the above equation, it can be seen that during the reaction, zinc is oxidized to zinc ions and hydrogen ions are reduced to hydrogen, a gas. Equation 2.2 can be divided into two reactions, the oxidation of zinc and the reduction of hydrogen ions:

Oxidation (anodic reaction) $Zn \rightarrow Zn^{+2} + 2e$ 2.3

Reduction (cathodic reaction) $2H + 2e \rightarrow H_2$ 2.4

An oxidation reaction or anodic reaction is indicated by an increase in valance or a production of electrons. A decrease in valence charge of the consumption of electrons signifies a reduction or cathodic reaction. Both reactions occur simultaneously and at the same rate. The oxidation of zinc is an electrochemical process [4], and it is an example of an electrochemical corrosion of zinc. The oxidation of aluminum proceeds by the same process:

$$2Al + 6HCl \rightarrow 2AlCl_3 + 3H_2$$ 2.5

Although this reaction appears to be different, both reactions involve the reduction of hydrogen ions and the production of hydrogen gas. Both reactions above involve the reduction of hydrogen ions and they differ only in their specific cathodic and anodic reactions.

2. Polarization. The concept of polarization is important for understanding corrosion behavior and corrosion reactions. The rate of an electrochemical reaction is limited by various physical and chemical factors. Therefore, an electrochemical reaction is said to be polarized or retarded by these environmental factors. Polarization can be divided into two different types, activation and concentration polarization.

Activation polarization refers to an electrochemical process that is controlled by the reaction sequence at the metal-electrolyte interface. This is easily illustrated by considering hydrogen gas evolution reaction on zinc (bare metal and not in a solution) during corrosion in acid solution. The following steps can be applied to the reduction of any species on a metal surface. The speed of reduction of the hydrogen ions will be controlled by the slowest step:

(a) The species must first be adsorbed (or attached) on the metal surface.
(b) The electron transfer must occur, resulting in a reduction of the species, (e.g., hydrogen).
(c) Two hydrogen atoms then combine to form a hydrogen molecule. These molecules then combine to form a bubble of hydrogen gas.

Concentration polarization refers to electrochemical reactions that are controlled by the diffusion of the electrolyte (solution of HCl) on the zinc metal such as in an aqueous solution. In the same reaction above, the number of hydrogen ions in solution is small, and the reduction rate is controlled by the diffusion of hydrogen ions to the metal surface. The reduction rate is controlled by processes (migration of ions, etc.) occurring within the bulk solution rather than at the metal surface in the solution.

Activation polarization usually is the controlling factor during corrosion in media containing a high concentration of active species (e.g., concentrated acids). Concentration polarization generally dominates the reaction when concentration of the reducible species is small (e.g., dilute acids).

3. Passivity. Passivity refers to the loss of chemical reactivity experienced by certain metals and alloys under particular environment conditions. In other words, certain metals and alloys become essentially inert and act as if they were noble metals as platinum and gold. It is important to understand that during the transition from active to the passive state, a 10^3 to 10^5 reduction in corrosion rate is usually observed. Passification is usually activation polarization due to the formation of a chemical film or protective barrier that is stable over a range of oxidizing power and is eventually destroyed in strong oxidizing solutions. An example of passification is iron in dilute alkali solution [5].

2.2.2 Environmental Aspects

Environmental factors affect corrosion rates, and some of the more common environmental variables are considered.

1. Effect of oxygen and oxidizers. The effect of oxidizers and oxidizing power was discussed above in connection with the behavior of active-passive metals. The effect of oxidizers on corrosion rate can be divided into different sections, and each provides a characteristic trend in the relationship of Corrosion Rate vs. Oxidizer Addition (referred to below as the plot). Reference will be made to this relationship in the following discussion.

   (a) Typical metals and active state of active-passive metals exhibit a linear relationship in the referenced plot. An example is iron in aerated water: $Fe + O_2 + H_2O$.
   (b) For metals which demonstrate active-passive transition, passivity is achieved only is a sufficient quantity of oxidizer of a sufficiently powerful oxidizer is added to the medium. An increase in corrosion rate followed by a rapid decrease, and then a corrosion rate which is essentially independent of oxidizer concentration, is characteristic of such active-passive metals and alloys such as 18Cr-8Ni stainless steel and titanium.
   (c) If an active-passive metal is initially passive in a corrosive medium, the addition of further oxidizing agents has only a negligible effect on corrosion rate. This condition frequently occurs when an active-passive metal is immersed in an oxidizing medium such as nitric acid or ferric chloride.

   The effect of oxidizer additions or the presence of oxygen on corrosion rate depends on both the medium and the metals involved. The corrosion rate may be increased by the addition of oxidizers, but they may have no effect on the corrosion rate, or a very complex behavior may be observed. By knowing the basic characteristics of a metal or alloy and the environment to which it is exposed, it is possible to predict in many instances the effect of oxidizer additions.

2. Effects of velocity. The effects of velocity on corrosion rate are, like the effects of oxidizer additions, complex and depend on the characteristics of the metal and the environment to which it is exposed. For corrosion processes, which are controlled by activation polarization, agitation and velocity have no effect on the corrosion rate. If the corrosion process is under cathodic diffusion control, then agitation increases the corrosion rate. The effect generally occurs when an oxidizer is present in very small amounts as in the case for dissolved oxygen in acids or water.

    If the process is under diffusion control and the metal is readily passivated, then the metal will undergo and active-to-passive transition. Easily passivated materials such as stainless steel and titanium frequently are more corrosion resistant when the velocity of the corrosion medium is high.

    Some metals owe their corrosion resistance in certain mediums to the formation of massive bulk protective films on their surface. These films differ from the unusual passivating films in that they are readily visible and much less tenacious. It is believed that both lead and steel are protected from attack in sulfuric acid by insoluble sulfate films. When materials such as these are exposed to extremely high corrosive velocities, mechanical damage or removal of these films can occur, resulting in accelerated attack by revealing fresh reactive surface the metal. This is called erosion corrosion, and the surface of aircraft aluminum is subject to this type of corrosion from exposure to very high velocity air, rain pelt and acids in the atmosphere.

3. Temperature. Temperature increases the rate of almost all chemical reactions, approximated as: 2 x the reactions rate for each 10 °C temperature increase. A rise in temperature explains the exponential increase in chemical reactivity, which translates to an increase in corrosion rate of most metals than can even affect passivated metal surfaces. By the same token, the corrosion rate can be decreased by lowering the temperature.

4. Effects of corrosion concentration. Many materials that exhibit passivity effects are only negligibly affected by wide changes in corrosive concentration. Other materials show similar behavior except at very high corrosive concentrations, when the corrosion rate increases rapidly. Lead is a material which show this effect, and it believed to be due to the fact that lead sulfate, which forms a protective film in low concentrations of sulfuric acid, is soluble in soluble in concentrated sulfuric acid. The behavior of acids which are soluble in all concentrations of water often increase the corrosion rate of metals rapidly at first followed by a decrease in corrosion rate. (i.e., as the concentration of corrosive is increased, the corrosion rate is likewise increased). This is primarily due to the fact that the amounts of hydrogen ions, which are the active species, are increased as acid concentration is increased. However, as acid concentration is increased further, corrosion rate reaches a maximum, and then decreases. This is due to the fact that at very

high concentrations of acids ionization is reduced. Because of this phenomenon, many of the common acids such as sulfuric, acetic, hydrofluoric, etc., are virtually inert when in the pure state, or 100% concentrations, and at moderate temperatures.

5. Galvanic coupling. In many practical applications, the contact of dissimilar materials is unavoidable. In Complex process streams and piping arrangements, different metals and alloys are frequently in contact with each other and the corrosive medium. Steel fasteners in aluminum aircraft skin are an example. The most active metal in a galvanic coupling will react fastest due to the greatest oxidation potential. The effect of galvanic coupling is usually to add an oxidizer to a corrosive solution (i.e., the rate of electron consumption is increased and the rate of metal dissolution increases). The effect of joining two dissimilar metals is generally to increase the corrosion rate of the cathodic reaction, but not always.

A classic example of galvanic coupling corrosion is coupled platinum and zinc metals immersed in dilute hydrochloric acid. Since platinum is inert in this medium, it tends to increase the surface at which hydrogen evolution can occur. Further, hydrogen evolution occurs more readily on the surface of platinum than on zinc. These two factors increase the rate of the cathodic reaction and consequently the corrosion rate of the zinc metal. The rate of the corrosion of the zinc is increased due to the galvanic coupling to less reactive metal, platinum.

6. Corrosion from biological organisms. Microorganisms are often the culprits responsible for the initiation of corrosion in crevices, joints and bonds that provide a moist breeding place for fungi, etc. Many species of fungi, mildew, bacteria and other organisms that can metabolize and reproduce on the surfaces of hardware are capable of producing corrosive by-products of metabolism including acids and sulfur gases (e.g., microorganism produced hydrogen sulfide can be converted to sulfuric acid via enzymes). Microorganisms have often, but not always, found to be responsible for the initiation of crevice corrosion.

2.2.3 Metallurgical Aspects

Metal and alloys are crystalline solids. The atoms of a metal are arranged in a regular, repeating array. Common crystalline arrangements of metals are: iron -- body-centered cubic structure, the austenitic stainless steels are face-centered cubic and magnesium -- a hexagonal close-packed lattice. Metallic properties differ from those of other crystalline solids such as ceramics and chemical salts. They are ductile and are good electrical and thermal conductors. These properties result from the nondirectional bonding of metals -- each atom is bonded to many of its neighbors, and the crystal structures are simple and closely packed.

Alloys are mixtures of two or more metals or elements. There are to kinds of alloys—homogenous and heterogeneous. Homogeneous alloys are solid solutions (i.e., the components are completely soluble in one another, and the materials has only one phase (e.g., 18–8 stainless steel). Heterogeneous alloys are mixtures of two or more separate phases that are not completely soluble in each other (e.g., low-carbon steel). Solid-solution or homogeneous alloys are usually more corrosion resistant than alloys with two or more phases, because galvanic couplings are present, but possess less strength and are more ductile.

2.3 Types of Corrosion

It is easy to classify corrosion by the forms in which it manifests itself, and the basis for this classification is the appearance of the corroded metal. Each form can be identified by mere visual observation. In most cases the naked eye is sufficient, but sometimes magnification is helpful or required. The following practical list is incomplete, but cover the majority of corrosion failures commonly observed on metals. The following discussion of corrosion is in terms of their characteristics, mechanisms and preventative measures.

The general types of corrosion that deteriorate metals include:

1. Uniform attack
2. Galvanic attack
3. Crevice corrosion
4. Pitting
5. Inter-granular corrosion
6. Selective leaching
7. Erosion corrosion
8. Stress corrosion

Each of the above types of corrosion is summarized below to give the reader a general understanding of where to expect corrosion and why it occurs.

2.3.1 Uniform Attack

Uniform attack is the most common form of corrosion. It is normally characterized by a chemical or electrochemical reaction that proceeds uniformly over the entire exposed surface or over a large area. The metal becomes thinner and eventually fails. For example, a piece of steel or zinc immersed in dilute sulfuric acid will normally dissolve at a uniform rate over its entire surface. Uniform attack, or general overall corrosion, represents the greatest destruction of metal on a tonnage basis. This form of corrosion, however, is not of too great concern from the technical standpoint, because the life of equipment can be accurately estimated on the basis of comparatively simple tests. Merely immersing specimens in the fluid involved is often sufficient. Uniform attack can be prevented or reduced by (1) proper materials, including coatings, (2) inhibitors, or (3) cathodic protection.

Most other forms of corrosion are considerable more difficult to protect. They are localized; attack is limited to specific parts of a structure. As a result, they tend to cause unexpected or premature failures. These are the corrosion failures usually observed in aircraft structures.

2.3.2 Galvanic or Two-Metal Corrosion

A potential difference usually exists between two dissimilar metals when they are immersed in a corrosive or conducive solution. If these metals are placed in contact (or otherwise electrically connected), this potential difference produces electron flow between them. Corrosion of the less corrosion-resistant metal (most chemically active) is usually increased and attack of the more resistant material is decreased, as compared with the behavior of these metals when they are not in contact. The less resistant metal become anodic (+) and the more resistant metal become cathodic (-). Usually the cathode (negative electrode where reduction occurs) or cathodic metal corrodes very little or not at all in this type of couple. Oxidation occurs at the positively charged anode that corrodes at a faster rate than the cathode. Because of the electric currents and dissimilar metals involved, this form of corrosion is called galvanic, or two-metal, corrosion.

The driving force for current and corrosion is the electrical potential developed between the two metals. Galvanic corrosion also occurs in the atmosphere. The severity depends largely on the type and amount of moisture present. Atmospheric exposure tests in different parts of the country have shown zinc to be anodic to steel in all cases, aluminum varied, tin and nickel always cathodic. Galvanic corrosion does not occur when the metals are completely dry since these are no electrolyte to carry current between the two electrode areas.

Accelerated corrosion due to galvanic effects is usually greatest near the junction, with attack decreasing with increasing distance from that point. Another important factor in galvanic corrosion is the area effect, or the ratio of the cathodic to anodic areas. An unfavorable area ratio (greater corrosion rate) consists of a large cathode and a small anode. For a given current flow in a cell, the current density is greater for a small electrode than for a larger one. The greater the current density at an anodic are the greater the corrosion rate. There are practical methods to reduce or eliminate galvanic corrosion.

2.3.3 Crevice Corrosion

Intense localized corrosion frequently occurs within crevices and other shielded areas on metal surfaces exposed to corrosives. This type of attack is usually associated with small volumes of stagnant solution caused by hoes, gasket surfaces, lap joints, surface deposits, and crevices under bolt and rivet heads. As a result, this form of corrosion is called crevice corrosion and sometimes deposit attack.

Examples of deposits, which may produce crevice corrosion, are sand, dirt, corrosion products and other solids. The deposits acts a shield and creates a stagnant condition thereunder. Oxygen could be reduced in crevices and could influence the rate of corrosion. Anerobic microorganisms, such as fungi, may be colonized in oxygen poor crevices and produced products of metabolism that may become corrosives (e.g., H_2O). Crevices serve as a place where deposits and moisture can initiate the incipient stages of corrosion.

Oxygen depletion has an important indirect influence, which become more pronounced with increasing exposure. After oxygen is depleted, no further oxygen reduction occurs, although the dissolution of metal such as aluminum continues. This tends to produce an excess of positive charge in a solution that is balanced by the migration of chloride ions in the crevice. This results in an increased concentration of metal chloride with the crevice. Metal (M) chlorides hydrolyze in water:

$$M^+ + H_2O = MOH \downarrow + H^+Cl^-$$ 2.6

The above equation shows that an aqueous solution of a typical metal chloride dissociates into insoluble hydroxide and a free acid. Chloride and hydrogen ions accelerate the dissolution rate of most metals and alloys. This increase in dissolution increases migration, and the result is a rapidly accelerating, or autocatalytic. process. The fluid within crevices exposed to neutral dilute sodium chloride solutions has been observed to contain 3–10 times as much chloride as the bulk solution and to possess a pH of 2–3. As the rate of corrosion within the crevice increases, the rate of oxygen reduction on adjacent surfaces also increases, and this cathodically protects the external surfaces. Thus, during crevice corrosion the attack is localized within shielded areas, while the remaining surface suffers little or no damage. Crevice corrosion often requires an incubation time of months to years before corrosion begins. Once it begins, it continues at an increasing rate.

Surfaces of metals such as aluminum with passive layers of oxide films are particularly susceptible to crevice corrosion. These films are destroyed by high concentrations of chloride or hydrogen ions.

Filiform corrosion is a special type of crevice corrosion. In most instances, it occurs under protective films, and for this reason is often referred to as underfilm corrosion. This type of corrosion is common. The most frequent example is the attack of enameled or lacquered surfaces of food and beverage cans that have been exposed to the atmosphere. The red-brown (iron rust) corrosion filaments are readily visible. Filiform corrosion has been observed on steel, magnesium and aluminum surfaces covered by tin, silver, gold, phosphate, enamel, and lacquer coatings. This type of corrosion does not weaken or destroy metallic components and only affects surface appearance although it could initiate more destruction corrosion. Under transparent surface films, the attack on coated steel cans appears as a network of corrosion product trails. The filaments (< 0.1 inch wide or less) consist of an active head and a red-brown corrosion product trail and corrosion occurs only in the filament head. The blue-green color of the active head is the characteristic color of ferrous ions, and the red-brown coloration of the inactive tail is due to the presence of ferric oxide or hydrated ferric oxide.

2.3.4 Pitting

A corrosion pit is a unique type of anodic reaction. It is an autocatalytic process—the corrosion processes within a pit produce conditions that are both stimulating and necessary for the continuing activity of the pit. An example of this process is a metal being pitted by an aerated sodium chloride solution. Rapid dissolution occurs within the pit, while oxygen reduction takes place on adjacent surfaces. This process is self-stimulating and self-propagating. The rapid dissolution of metal within the pit tends to produce and excess of positive charge in this area, resulting in the migration of chloride ions to maintain electroneutrality. Thus, in a pit there is a high concentration of MC1 and, as a result of hydrolysis, a high concentration of hydrogen ions. Both hydrogen and chloride ions promote the dissolution of most metal and alloys. Since the solubility of oxygen is virtually zero in concentrated solutions, no oxygen reduction occurs within a pit.

Measurements of pit depth are complicated by the fact that there is a statistical variation in the depths of pits on an exposed specimen. The average depth of pits is not so important because it is the deepest pits that cause failure.

2.3.5 Intergranular Corrosion

High strength aluminum alloys depend on precipitated phases for strengthening and are susceptible to intergranular corrosion. The separate phases within an alloy and adjacent materials provide potential differences that make them susceptible to chemical attack.

2.3.6 Selective Leaching

Selective leaching is the removal one element from a solid alloy by corrosion processes.

2.3.7. Erosion Corrosion

Velocity often strongly influences the mechanisms of the corrosion reactions. It exhibits mechanical wear effects at high values and particularly when the solution contains solids in suspension. he solids can serve to erode the metal (like sandpaper) and produce fresh unprotected metal. Increases in velocity generally result in increased attack, particularly if substantial rates of flow are involved. The effect may be nil or may increase slowly until a critical velocity is reached, and then the attack may increase at rapid rate. Erosion corrosion can occur on metals and alloys that are completely resistant to a particular environment at low velocities.

2.3.8 Stress Corrosion

Stress corrosion cracking refers to cracking caused by the simultaneous present of tensile stress and a specific corrosive medium. During stress corrosion cracking, the metal or alloy is virtually unattacked over most of its surface, while find cracks progress through it. This cracking phenomenon has serious consequences since it can occur at stresses within the range of typical design stress. An example is exposure of 304 stainless steel to boiling $MgCl_2$ at 310 °F (154 °C) that reduces the strength capability. Two more examples are the stress corrosion cracking of brass and the caustic embrittlement of steel.

Increasing the stress decreases the time before cracking occurs. This is some conjecture concerning the minimum stress required before cracking.

There are other types of corrosion [5] that would be rightfully included in a comprehensive discussion that are not a part of this book. Therefore, references are given to assist the reader in finding detail information on specific subjects involving corrosion.

3

Fundamentals of Electromagnetic Shielding

3.1 Definition of Shielding Effectiveness

The shielding effectiveness (SE) is typically defined as the ratio of the magnitude of the incident electric field, E_i, to the magnitude of the transmitted electric field, E_t:

$$SE = \left| \frac{\vec{E}_i}{\vec{E}_t} \right| \qquad SE(dB) = 20 \cdot \log_{10}\left(\frac{E_i}{E_t} \right) \qquad 3.1$$

The above definition of shielding effectiveness is sometimes referred to as the electric field shielding effectiveness (ESE) since it involves the ratio of electric field amplitudes. Shielding effectiveness can also be defined in terms of the ratio of the incident and transmitted magnetic field amplitudes, which is sometimes referred to the magnetic field shielding effectiveness (MSE). For plane waves and the same media (e.g., air) on both sides of the shield, these two definitions are equivalent (i.e., ESE = MSE).

3.2 Factors that Determine Shielding Effectiveness

There are several factors that determine the effectiveness of an electromagnetic shield. These factors include the following:

1. frequency of the incident electromagnetic field;
2. shield material parameters (conductivity, permeability, and permittivity);
3. shield thickness;
4. type of electromagnetic field source (plane wave, electric field, or magnetic field);

5. distance from the source to the shield;
6. shielding degradation caused by any shield apertures and penetrations; and
7. quality of the bond between metal shield surfaces.

3.3 Electromagnetic Shielding Theory

Electromagnetic fields, in a classical sense, are governed by Maxwell's equations subject to a set of boundary conditions. Schelkunoff [5], using a transmission line analogy, laid out the basic theory for plane wave attenuation through shields. The theory has subsequently been expanded upon by other authors [e.g., 6]. Using Schelkunoff's formulation (which assumes that the shield is located in the far field of the source), the plane wave shielding effectiveness (in dB) can be expressed as the sum of an absorption loss (A) and a reflection loss (R) plus a multiple reflection correction term (B):

$$SE = A + R + B \qquad\qquad 3.2$$

Electromagnetic shielding processes are shown pictorially in Fig. 3.1. A shield has two boundaries where reflection and transmission of electric (E) and magnetic (H) fields occur. Incident fields are depicted by "i" subscripts, reflected fields by "r" subscripts, and transmitted fields by "t" subscripts. At each air/metal interface, a portion of the field is reflected and the remainder of the field is transmitted. Some of the power is absorbed within the shield. Reflection occurs due to an impedance mismatch. Absorption occurs due to energy losses within the shield material. The boundary conditions, which require the tangential fields to be continuous at the interfaces, determine the magnitude of the reflection and transmission coefficients.

For a good shield, a large portion of the incident field (E_i) is reflected and only a small portion is transmitted (E_t). Thus, there is usually a large reflection loss (R) at the first air/shield interface. The transmitted portion experiences an absorption loss (A) before encountering the second shield/air interface. A small portion of this field (E_{tt}) is then transmitted through the shield to the other air medium. If the absorption loss is large, this transmitted portion determines the overall shielding effectiveness of the shield (i.e., $SE = A + R$). However, if the absorption loss is small, multiple reflections within the shield can cause a reduction in shielding effectiveness due to the presence of many significant higher order terms (i.e., E_{tr2t}, E_{tr4t}, E_{tr6t}, etc.) that add vectorially to the transmitted field (E_{tt}). This reduction in shielding effectiveness due to multiple reflections manifests itself as a large, negative B term in Eq. 3.2. The absorption, reflection, and multiple reflection correction terms in Eq. 3.2 are dependent upon the impedance of the incident electromagnetic field and/or the characteristics of the shield material as described below.

3.3.1 Absorption Loss

Absorption loss is incurred as the electromagnetic wave penetrates the shield. The amplitude of the electric and magnetic fields decay exponentially due to the induced currents and resultant ohmic losses in the material. The propagation constant of the fields in the metal shield is given by

$$\gamma_s = \sqrt{j\omega\mu_s\sigma_s}$$ 3.3

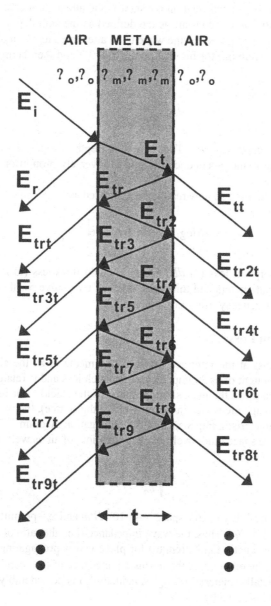

Fig. 3.1. Illustration of reflection and transmission of electromagnetic fields.

where: γ_s = propagation constant of wave inside shield, $j = \sqrt{-1}$, $\omega = 2\pi f$ = radian frequency (in rad/sec), f = frequency (in Hz), μ_s = shield permeability (in H/m), and σ_s = shield conductivity (in $mhos/m$).

The real part of this propagation constant is the attenuation constant and the reciprocal of the attenuation constant is the defined as the skin depth. Thus, the skin depth is the distance required for the wave to be attenuated to $1/e$ or about 37% of its initial amplitude. Solving for the real part of Eq. 3.3 and then taking the reciprocal results in

$$\delta_s = \sqrt{\frac{2}{\omega \mu_s \sigma_s}} \qquad 3.4$$

where δ_s = skin depth (in m). Thus, the skin depth is reduced by increasing the shield permeability and/or conductivity and is inversely proportional to the square root of frequency.

The absorption loss can now readily be expressed as:

$$A = 20 \log_{10}(e^{t_s/\delta_s}) = 8.686\left(\frac{t_s}{\delta_s}\right) \qquad 3.5$$

where: A = absorption loss (in dB) and t_s = shield thickness (in m). Knowing the shield material parameters and thickness, one can then solve for the absorption loss as a function of frequency using Eqs. 3.4 and 3.5.

3.3.2 Reflection Loss

The reflection loss at the interface between two media (e.g., the air and the metal shield) can be thought of as an impedance mismatch loss and is related to the ratio of the wave impedance to the intrinsic impedance of the shield. For plane waves (i.e., radiated electromagnetic waves in the far-field of the source), the wave impedance is equal to the characteristic impedance of the propagation medium. The characteristic impedance of free space (and, in close approximation, of air as well) is given by

$$Z_0 = \sqrt{\frac{\mu_0}{\varepsilon_0}} = 377\Omega \qquad 3.6$$

where: μ_0 = permeability of free space = $4\pi \cdot 10^{-7}$ H/m and ε_0 = permittivity of free space = $8.854 \cdot 10^{-12}$ F/m. Thus, the wave impedance (i.e., the ratio of the electric field strength to the magnetic field strength) for plane waves propagating in free space is also 377 Ω. On the other hand, the intrinsic impedance of the shield material (which is typically a metallic material of high conductivity) is considerably lower than that of free space and is given by

$$Z_s = \sqrt{\frac{j\omega\mu_s}{\sigma_s}} \qquad\qquad 3.7$$

where Z_s = intrinsic impedance of metal shield (in Ω).

Because of the large impedance mismatch, a large portion of the incident field is reflected at the air/shield interface and only a small portion is transmitted. The transmission coefficient at the first air/shield interface is given by

$$T_1 = \frac{E_t}{E_i} = \frac{2Z_s}{Z_o + Z_s} \qquad\qquad 3.8$$

Similarly, the transmission coefficient at the second shield/air interface is given by

$$T_2 = \frac{E_{tt}}{E_t} = \frac{2Z_o}{Z_o + Z_s} \qquad\qquad 3.9$$

The combined transmission coefficient is given by the product of Eqs. 3.8 and 3.9

$$T_{tot} = T_1 T_2 = \frac{E_{tt}}{E_i} = \frac{4Z_o Z_s}{(Z_o + Z_s)^2} \qquad\qquad 3.10$$

The reflection loss is thus defined as

$$R = -20\log_{10}|T_{tot}| = 20\log_{10}\frac{|1+k|^2}{4|k|} \qquad\qquad 3.11$$

where $k = Z_s/Z_o$. Knowing the shield material parameters and the frequency, one can then solve for the reflection loss using Eqs. 3.6, 3.7, and 3.11. For low impedance shields, k is very small which results in a large reflection loss.

3.3.3 Multiple Reflection Correction Term

The multiple reflection correction term accounts for the reduction in shielding effectiveness for shields with low absorption loss due to multiple reflections inside the shield. This term is given by

$$B = 20\log_{10}\left|1 - \frac{(k-1)^2}{(k+1)^2}e^{\frac{-2t_s}{\delta_s}}\right| \qquad\qquad 3.12$$

The total shielding effectiveness of a shield can now be calculated from Eqs. 3.2, 3.5, 3.11, and 3.12.

3.3.4 Effect of Near-Field Sources on Shielding Effectiveness

As stated previously, the Schelkunoff formulation assumes that the shield is located in the far field of the source. Thus, the incident field is assumed to be a plane wave with an impedance equal to that of free space ($Z_o = 377 \ \Omega$). If the shield is located in the near-field of an emitter, the R and B equations (Eqs. 3.11 and 3.12, respectively) must be slightly modified to correct for a wave impedance, Z_w, that differs from that of free space. Thus, for near-field illumination of a shield, the parameter k in Eqs. 3.11 and 3.12 is replaced by a modified parameter k' given by

$$k' = \frac{Z_s}{Z_w} \qquad\qquad 3.13$$

For electric field sources, the wave impedance is typically assumed to be that of an electric dipole:

$$Z_{we} = Z_o \frac{j/(\beta_o r) + 1/(\beta_o r)^2 - j/(\beta_o r)^3}{j/(\beta_o r) + 1/(\beta_o r)^2} \qquad\qquad 3.14$$

where $\beta_o = 2\pi/\lambda$, λ = wavelength, and r = distance from the source. Similarly, for magnetic field sources, the wave impedance is typically assumed to be that of a magnetic dipole (electrically small loop):

$$Z_{wm} = Z_o \frac{j/(\beta_o r) + 1/(\beta_o r)^2}{j/(\beta_o r) + 1/(\beta_o r)^2 - j/(\beta_o r)^3} \qquad\qquad 3.15$$

In the extreme near field ($r \ll \lambda/2\pi$), Eqs. 3.14 and 3.15 simplify to

$$Z_{we} \approx \frac{Z_o}{j\beta_o r} \qquad\qquad 3.16$$

$$Z_{wm} \approx jZ_o\beta_o r \qquad\qquad 3.17$$

Figure 3.2 shows the wave impedance versus distance for far-field, electric, and magnetic field sources at a frequency of 1 *MHz*.

3.4 Practical Electromagnetic Shielding

Since the conductivity of commonly used metallic shield materials, such as aluminum, is very high (typically on the order of 10^7 *mhos/m* or higher), a metallic barrier theoretically provides a large amount of attenuation. As a rule of thumb, if a continuous metallic barrier has a thickness sufficient to support itself (say 30 *mils* or more), the theoretical shield effectiveness is extremely high (typically on the order of 100 *dB* or more) at radio frequencies (RF).

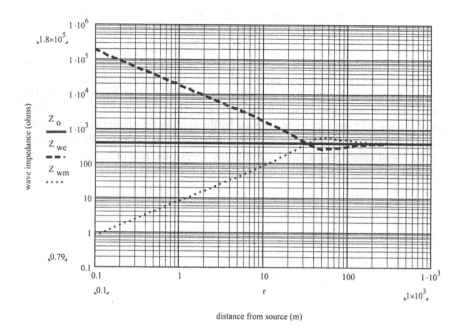

Fig. 3.2. Wave impedance vs. distance from source for plane wave, electric field, and magnetic field sources (f = 1 MHz)

In practice, however, the shielding barrier is not continuous but instead contains seams that are fastened (bonded) together with metallic fasteners. Often, the joints allow electromagnetic fields to leak through the shield thereby degrading the overall shielding effectiveness. The quality of the electrical bond across the faying surfaces at RF frequencies thus, in large part, determines the overall shielding effectiveness of the shielding structure. For example, a shielding effectiveness on the order of 20 *dB* is typical for riveted aluminum panels with nonconductive sealants.

Providing a direct current (dc) path across the faying surface is necessary, but not sufficient, for high shielding effectiveness (riveted joints, in fact, often provide a very low dc resistance between bonded panels despite their rather poor shield effectiveness). Rather, one must provide a low RF impedance path across the faying surfaces. RF surface currents flowing across a joint in the skin of an aircraft induce electric fields in the interior of the aircraft. These fields in turn induce currents and voltages on interior wiring. The maximum voltage that can be induced on an interior wire is the voltage drop across the joint given by

$$V_j = J_s Z_{tj} \qquad\qquad 3.18$$

where: V_j = voltage drop across the joint (in *V*), J_s = surface current density (in *A/m*), and Z_{tj} = transfer impedance of the joint (in *Ωm*). Thus, a good measure of how well a joint is bonded is the transfer impedance of the bonded joint across the frequency band of interest [8].

3.5 Role of Conductive Sealants in Electromagnetic Shielding

Low transfer impedance is achieved by a bond that provides continuous, uniform conductivity between the faying surfaces. The application of conductive sealants in the faying surfaces of the panels can significantly reduce the transfer impedance of the joint thereby reducing the penetration of electromagnetic fields through the joints. Thus, in large part, a significant portion of the inherent shielding effectiveness of the metallic panels can be retained by the application of conductive sealants. Selecting conductive sealants that are chemically stable and galvanically compatible with the metal substrates can also significantly reduce the rate of corrosion thereby maintaining an effective electrical and mechanical bond.

4

Investigation of the Relationship Between DC Resistance and Shielding Effectiveness

4.1 Determining a Relationship Between DC Resistance and Shielding Effectiveness

An essential ingredient of EMP hardening, particularly for airborne systems, is electromagnetic shielding. In practice, shields are rarely formed of single, continuous sheets of metal. Sheets must be joined together to form the shield, and the joints must be properly bonded to provide desirable shielding performance of a broad frequency range. Electromagnetic shielding effectiveness was, therefore, selected as the measure of bond effectiveness with regard to EMI/EMP hardness. A stainless steel test joint was fabricated, and the electromagnetic shielding effectiveness of the test joint was measured as a function of dc resistance. The stainless steel test joint, dc resistance measurements, shielding effectiveness measurements, and the measurement results are described below.

4.2 Stainless Steel Test Joint for DC Resistance and Shielding Effectiveness Measurements

In order to determine the relationship, if any, between dc resistance and shielding effectiveness, s specialized test joint was designed and fabricated. The test joint was constructed out of No. 316 stainless steel rather than aluminum because of the superior ability of stainless steel to withstand atmospheric corrosion. A sketch of the stainless steel test joint is shown in Fig. 4.1. The back plate of the test joint was designed to conform with the flange area of the shielding effectiveness test box. Dielectric washers were used to insulate the screw fasteners from the top metal plate and to allow a fixed separation to be maintained between the plates. The use of dielectric washers allowed the resistance between the plates to be controlled by the bulk conductivity of a conductive sealant material that was applied in the 3/4 inch overlap area between the plates. Cylindrical metal posts were welded onto each plate to facilitate the dc resistance measurements.

A front view of the stainless steel test joint is shown in Fig. 4.2. The two overlapping plates connected with screw fasteners are visible as are the two metal posts used for the resistance measurements. A side view of the stainless steel test joint is shown in Fig. 4.3. During resistance and shielding effectiveness tests, the air gap (visible in the side view of Fig. 4.3) was filled with conductive sealant of varying resistivity. A rear view of the stainless steel test joint after loading with conductive sealant is shown in Fig. 4.4.

Conductive sealants were formulated to yield a wide range of resistance values above and below 2.5 mΩ. These materials consisted of silver powder dispersed in urethane resins at different volume loadings to vary the resistivity. One sealant was a two-part urethane, whereas the other four sealant types were moisture-cured urethanes. Each sealant was applied to the stainless steel test joint, cured as required, and subsequently measured for electrical resistance and shielding effectiveness. (These sealant types are not intended for actual use on aircraft, as silver and aluminum will corrode in the presence of moisture).

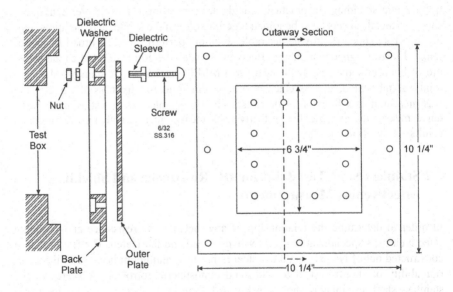

Fig. 4.1. Sketch of stainless steel test joint.

Fig. 4.2. Front view of stainless steel test joint.

Fig. 4.3. Side view of stainless steel test joint.

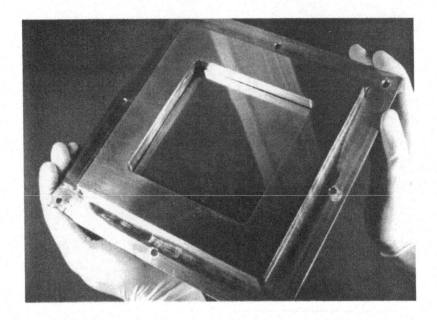

Fig. 4.4. Rear view of stainless steel test joint.

4.3 DC Resistance Measurements

The dc resistance values of the various conductive sealant test samples were measured with a double Kelvin bridge milliohmmeter. The double Kelvin bridge is a four terminal instrument which is capable of accurately measuring very small resistance values (in this case, as low as 0.1 mΩ). For resistance values less than 1 Ω, the terminals of the milliohmmeter were clipped or contacted directly to the metal plates of the test joint in order to remove the additional resistance associated with the bonded posts from the measurement. All resistance measurements were made after the sealant material had cured so that the resistance value did not vary appreciably with time. The shielding effectiveness tests were run immediately after the resistance measurements were completed.

4.4 Shielding Effectiveness Measurements

The shielding effectiveness of the test joint samples was evaluated as a measure of the EMI/EMP hardness of the bonded joint. The shielding effectiveness of each test sample was determined as the difference (in dB) between the electromagnetic field coupled through an aperture with and without the sample present over the aperture. (This measurement technique conforms to the basic procedure set forth in MIL-STD-285[2].) A block diagram of the test setup used to make the shielding effectiveness measurements is illustrated in Fig. 4.5. This test setup enabled accurate and repeatable swept frequency measurements to be made on the test joint samples over the 500–1000-MHz frequency range.

A transmitting antenna (capacitively-loaded dipole) was placed in the center of a shielding effectiveness test enclosure, and receiving antenna (log conical spiral) was positioned outside of the test enclosure at a distance (5-1/2 feet), which meets the far-field criterion over the entire test frequency range. The log conical spiral receiving antenna was directed towards a 6 inch square aperture located in one wall of the enclosure. The purpose of the test chamber was to isolate the transmitting and receiving antennas so that the only coupling between them was through the aperture. To minimize unwanted coupling, all permanent seams of the test enclosure were welded, and finger stock was installed around the lid of the enclosure and around the periphery of the test panel port (aperture). The inside of the test enclosure was lined with ferrite absorbing tiles which served to dampen enclosure resonances and to simulate free space conditions for the transmitting antenna.

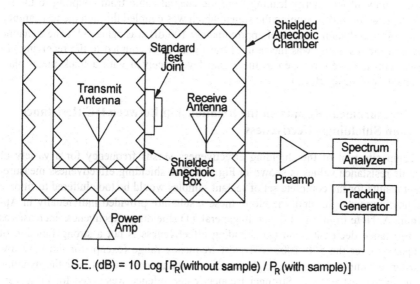

S.E. (dB) = 10 Log [P_R(without sample) / P_R (with sample)]

Fig. 4.5. Block diagram of the shielding effectiveness test setup.

The test joint was fastened to the test enclosure and electromagnetically sealed by three rows of finger stock located on the enclosure flange area. Conductive copper tape was also applied around the periphery of the test sample as an additional precautionary measure to prevent the electromagnetic field from leaking round, rather than through, the test sample. A 6-dB attenuator was used at the transmitting antenna input port to ensure a stable 50 Ω impedance at the power amplifier output. A tracking generator was used in conjunction with a spectrum analyzer to provide the swept frequency output source and receiver. (The tracking generator derives its output signal from the first and third local oscillator outputs of the spectrum analyzer.) A 10-dB pad was used to attenuate the nominal O dBm output signal level of the tracking generator to approximately -10 dBm at the per amplifier input. The power amplifier provides approximately 46 dB of RF gain which corresponds to 4 watts output power with -10 dBm input power. A low noise, high gain amplifier (noise figure = 2 dB; gain = 35 dB) was used to improve the sensitivity of the spectrum analyzer and thereby increase the overall dynamic range.

To isolate the test setup from the external electromagnetic environment and to minimize errors due to ground and/or wall reflections, the test enclosure and receiving antenna were located in a shield anechoic chamber. The isolation characteristics of the chamber were 100 dB or greater over the frequency range of 10 MHz to 1000 MHz and absorption was effective for RF frequencies greater than 300 MHz. The test enclosure was connected to the outside of the anechoic chamber via a 1/2 inch copper pipe/flange assembly, and a coaxial cable routed through the copper pipe connected the power amplifier outside the enclosure to the attenuator and transmitting antenna located inside the test enclosure. This cable routing configuration was used to prevent RF energy leaking from the coaxial cable from coupling to the receiving antenna and swamping the signal which was coupled through the test sample. In this way, a dynamic range of approximately 100 dB was achieved. (The dynamic range of the test setup was the ratio of the signal level coupled to the receiving antenna with the aperture open to the signal level received with a thick metal plate fastened over the aperture.)

4.5 Measurement Results on the Relationship Between DC Resistance and Shielding Effectiveness

A composite plot of the shielding effectiveness versus frequency for a variety of sample resistance values is shown in Fig. 4.6. The shielding effectiveness measurement results for the complete set of sealant samples would be too cluttered to fit on a single composite plot and, therefore, these results are provided individually in Appendix A. Note from Fig. 4.6 that, in general, (1) shielding effectiveness increases as dc resistance decreases, and (2) shielding effectiveness is not a strong function of frequency over the 500–1000 MHz test frequency range (except for the very low resistance samples which has high shielding effectiveness values near the dynamic range of the test system). Stronger frequency dependence was noted for these very low resistance samples due to frequency dependent leakage paths, which are signifi-

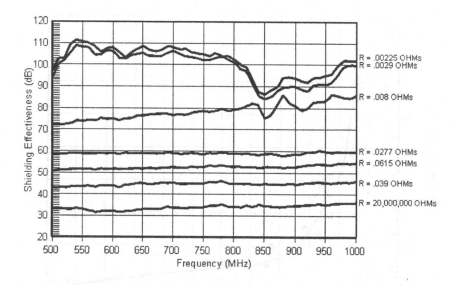

Fig. 4.6. Shielding effectiveness vs. frequency for different bond resistances.

cant at the higher shielding effectiveness values but insignificant at the lower shielding values.

In order to summarize the findings on a simple 2-dimensional plot, the shielding effectiveness versus frequency curves were reduced to an average shielding effectiveness value in dB (i.e., the frequency average of the logarithmic shielding values) and then all of the average values were plotted on a graph of shielding effectiveness versus dc resistance. The resulting plot of the (average) shielding effectiveness versus dc resistance is illustrated in Fig. 4.7. First, note that there appears to be a limiting shielding effectiveness value for the high resistance samples which corresponds to a minimum shielding effectiveness of approximately 35 dB. This "minimum" shielding effectiveness value is a function of the test frequency range, the electromagnetic wave impedance and, as will be later demonstrated, the particular test joint geometry used. Second, note that there also appears to be a limiting shielding effectiveness value for the low resistance samples which corresponds to a maximum shielding effectiveness of approximately 105 dB. Closer scrutiny of the data indicates that this "maximum" shielding effectiveness value in some instances occurs due to a limitation in the test sample (i.e., RF leakage) but in other instances results from the limited dynamic range of the test system. And, finally, note the sharp decrease in shielding effectiveness from approximately 105 dB to approximately 55 dB as the dc resistance of the test sample increases from approximately 2.5 mΩ to approximately 100 mΩ. Although the test results presented here represent only one test joint geometry and one electromagnetic test environment, the data of Fig. 4.7 strongly suggest that relaxation of the 2.5 mΩ requirement could result in a significant reduction of the EMI/EMP hardness of aircraft and weapons systems.

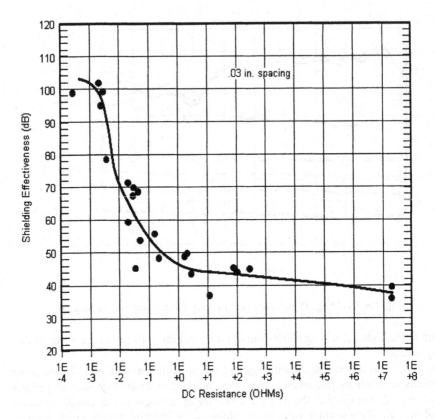

Fig. 4.7. Shielding effectiveness vs. dc resistance.

It should be emphasized that the data presented here by no means is intended to imply that a low dc resistance is the only requirement for a quality bond from and EMI/EMP hardness perspective. To demonstrate this point, the test joint was modified by removing the insulating washers to allow the screw fasteners to establish dc electrical continuity between the metal plates. The resulting dc resistance of the test sample without a conductive sealant in the flange area was extremely low (<0.1 mΩ) whereas, as can be seen from Fig. 4.8, the shielding effectiveness was generally very poor and frequency sensitive. This example clearly shows that a low dc resistance value is not sufficient for good shielding effectiveness and that electrical continuity must be obtained across the entire joint area (resulting in low RF bonding impedance) for effective shielding over a broad frequency range. This requirement is not necessarily met by a low dc resistance value not can it be verified solely by a dc resistance measurement.

4.6 Effect of Plate Separation on Shielding Effectiveness

A brief experiment was conducted to demonstrate the influence of the test joint geometry on the shielding effectiveness test results. The experiment consisted of measuring the shielding effectiveness of the test joint without conductive sealants while varying the separation distance between the metal plates. Fixed separation distances were obtained by placing one or more Mylar® sheets (minimum thickness = 0.1 mm) in the flange area between the plates. The tests were run with 0.1, 0.3, 0.95 and 2.5-mm plate separation distances, respectively.

The shielding effectiveness versus frequency test results for each plate separation distance are provided in Appendix B. Fig. 4.9 shows a composite plot of the average (over frequency) shielding effectiveness values for the various plate separation distances. Note that the shielding effectiveness is inversely proportional to the plate separation distance. Approximately 15 dB decrease in shielding effectiveness was measured per decade increase in plate separation distance over the range of plate separation distances evaluated. (Intuitively, a 20 dB per decade roll off could be expected from the decreasing capacitive reactance between the plates (neglecting fringing terms). Fringing terms could reduce the slope of the roll off somewhat since these terms become more significant at greater separation distances.) Also note that the data from Fig. 4.6 was generated with a dielectric spacer around each screw fastener, which maintained a fixed 30 *mil* spacing between the plates. The limiting value for the high resistance test samples was approximately 35 dB, which falls very near the curve shown in Fig. 4.9 for a 0.76 mm (30 mil) plate separation distance.

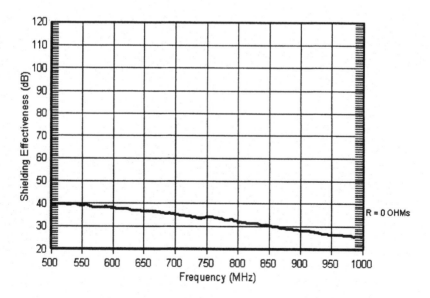

Fig 4.8. Shielding effectiveness vs. frequency for test sample with plates short-circuited together through screw fasteners but having 2.5 mm plate separation.

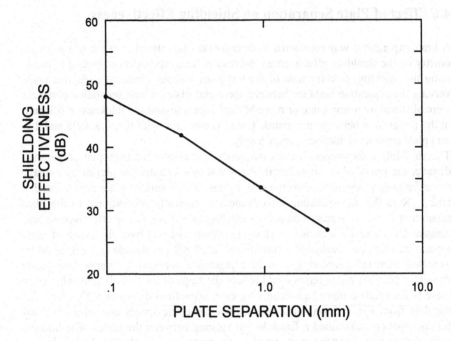

Fig. 4.9. Shielding effectiveness vs. plate separation distance.

5

Identification and Evaluation of Optimum Conductive Sealant Materials

5.1 Identification of Materials

The objective of this task was to identify and evaluate optimum corrosion prevention materials for use on nuclear hardened aircraft and weapon systems. New test joints were designed and fabricated out of 7075 aluminum and fastened using realistic torque values. The aluminum test joints were loaded with the top candidate sealant materials and weathered in a salt fog environment according to ASTM B117. The dc resistance and shielding effectiveness of the test joints were measured before weathering tests began and periodically during the weathering tests to monitor the corrosion effects on the electrical and electromagnetic performance of the bonds. In addition, control joints (no sealant) were weathered and tested simultaneously to compare the relative merit of the selected sealant materials.

A variety of conductive sealant materials were considered for use as optimum compromises between EMI/EMP hardness and corrosion prevention. The use of metal-coated glass spheres was considered early in the program. However, it was later learned that these materials are not suitable for aircraft structural type sealants for the following reasons.

1. The number of contact points between spheres is small compared to that for irregular shaped particles. The greater the number of contact points, the less damage likely to be experienced from vibration.
2. During normal stresses from aircraft, vibration can cause shearing through the thin metal coating on the spheres, which will destroy the current path.
3. Temporary current overload or pulsing, such as that induced by lightning or nuclear EMP, can damage the coating (as shown in Fig. 5.1).

Pure silver powders and flakes are highly desirable conductive particles from an electrical performance point of view for use in EMI gaskets, compounds, coatings and adhesives. However, the metal's relatively high density, constantly fluctuating cost and potential for galvanic corrosion with many substrate materials often prohibit its use. To overcome these problems, the EMI industry has developed hybrid particles that offer the beneficial conductivity characteristics of silver, yet minimize or negate the drawbacks concerning density, cost and galvanic corrosion. For aluminum substrates, a particularly promising material that was identified was silver-coated aluminum. This hybrid material, produced by coating a thin layer of silver on the surface of aluminum particles, has a galvanic potential very near that of aluminum (see Table 5.1) and, thus, was expected to have superior corrosion characteristics when used on aluminum substrates.

The final list of conductive sealant materials, which appeared to be most promising for meeting the program objectives, is shown in Table 5.2. These six conductive sealant materials were therefore selected for evaluation under accelerated environmental test conditions. A salt fog exposure environment per ASTM B117 was selected to simulate accelerated weather conditions. The dc resistance and shielding effectiveness of the test joints and the control joints were measured before weathering tests began and periodically during the weathering tests to determine the relative degradation in electrical performance as a function of exposure time in the salt fog environment.

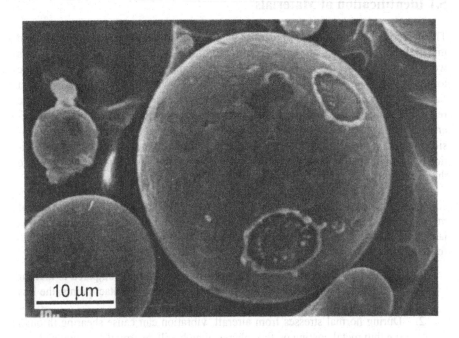

Fig. 5.1. Silver-coated glass spheres and damage to coating from current overload.

Table 5.1. Corrosion Potentials of Various Metals and EMI Gasket Materials (In 5% NaCl Solution at 21°C After 15 Minutes of Immersion.)

Material	E_{CORR} Vs. SCE[a] (Millivolts)
Pure Silver	-25
Silver-Filled Elastomer	-50
Monel Mesh	-125
Silver-Plated-Copper Filled Elastomer	-190
Copper	-244
Tin-Plated Berylliumm-Copper	-440
Tin-Plated Copper-Clad Steel Mesh	-440
Aluminum[b] (1000)	-730
Silver-Plate-Aluminum Filled Elastomer	-740

[a]Galvanic Potential of Test Cell Measured with Standard Calomel Electrode as a Reference.
[b]Aluminum Alloys Approximate =700 to –840 mV vs. SCE in 3% NaCl. Mansfield F. and Kenkel, J.V., laboratory Studies of Galvanic Corrosion of Aluminum alloys, "Galvanic and Pitting Corrosion –Field and Lab Studies, ASTM STP 576, 1976, pp. 20-47.

Table 5.2. Conductive Sealant Materials Selected for Testing.

Conductive Sealant	Manufacturer
1. Polysilicone RTV # 1038, Ag/Al[a] Filler	Chomerics, Inc
2. Polysilicone RTV # 1038, Al[b] Filler	Chomerics, Inc.
3. Polyurethane RTV # 4329-25-1, Ag/Al Filler	Chomerics, Inc.
4. Polyurethane RTV # 4329-25-2, Al Filler	Chomerics, Inc.
5. Proseal #872 Polysulfide, Al Filler	Products Research Corp.
6. Proseal RW2-28-71 Al/Ni	Products Research Corp.

[a]Silver coated aluminum powder
[b]Aluminum powder

Fig. 5.2.. Test joint base plate (rear side) and cover, polished to remove oxides and surface defects, and cleaned with methanol. The aluminum flat sheet #7065-76 was used for fabrication of each joint. The aluminum flat sheet was certified per Federal Specification QQ-A-250/12.

5.2 Evaluation of Conductive Sealants in Salt Spray Environment per ASTM B117

DC resistance measurements were performed on the conductive sealants shown in Table 5.2 at 24-hour time intervals during the salt fog exposure testing [7]. Shielding effectiveness measurements [8] were made on the aluminum- and nickel-filled poly-sulfide sealant (Products Research Proseal RW2-28-71) and the aluminum-filled polyurethane RTV sealant (Chomerics #43429-252) at approximately one week time intervals during the accelerated environmental exposure testing. A description of the aluminum test joints and the results of the dc resistance and shielding effectiveness measurements are presented below.

5.2.1 Aluminum Test Joints

Aluminum test joints were fabricated for use in the weathering study. The test joints were designed and constructed to fit on the shielding effectiveness test enclosure while being representative of actual joints located on the skin of military aircraft and missile systems. The material chosen for the test joints was standard aircraft 7075, T6 aluminum certified per Federal Specification QQ-A-250/12. The 7075 aluminum was chosen because it is a common aircraft type of aluminum and is most prone to

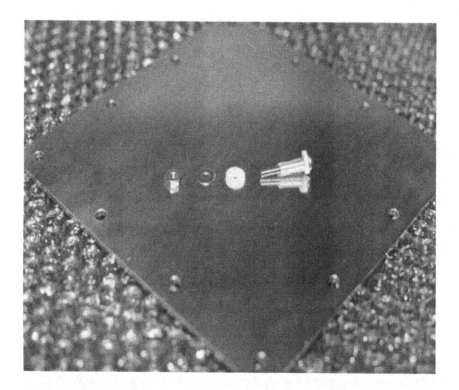

Fig. 5.3 Fastener assembly: (1) Stainless steel #316 6/32 round slotted screw insulated with #6 stainless steel washer and (4) hex-type stainless steel nut. Each assembly was torqued to 13 inch-lbs.

corrosion compared to other types of aluminum. The design of the aluminum test joints was identical to that of the stainless steel test joint with two exceptions:

1. Rather than being threaded, the back plate had countersunk holes to allow screw/nut fasteners (described in MIL-STD-QQS267) to be used, and
2. No dielectric spacers were used between the plates.

The cover and base plates of a fabricated joint assembly are shown in Fig. 5.2. The screw fasteners (see Fig. 5.3) are electrically insulated from the test joint to enable the metal-to-metal joint to establish the electrical conductivity and to prevent a galvanic cell from being created between the stainless steel fasteners and the aluminum plates. The conductive sealant was applied to the base plate of the test joint as shown in Fig. 5.4. The cover plate was than fastened to the base plate with the screw fastener assembly and torqued to 13 inch-lbs using a torque wrench, as illustrated in Fig. 5.5. The insulation of the screw fasteners was tested by verifying with an ohmmeter than an "open" connection was measured between the screws and the cover plate and between the nuts and the base plate, as illustrated in Fig. 5.6.

Fig. 5.4. Application of conductive selant (#1038 Ag/Al in RTV polysilicone) applied to base plate of joint (front side). The sealants are typically viscous and require troweling to cover the joint area

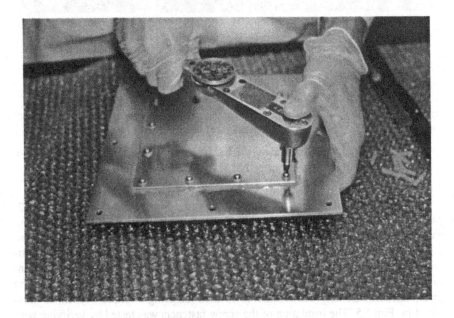

Fig. 5.5. Fastener assembly torqued to 13 inch-lbs using a torque wrench.

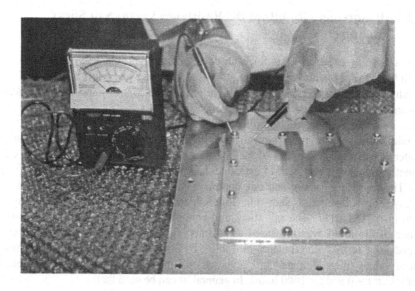

Fig. 5.6. Testing of insulation fastener assembly with ohm-meter.

Fig. 5.7. Resistance measurement of test joint using a General Electric double Kelvin bridge millohm-meter.

5.2.2 Measurement Results of DC Resistance Versus Salt Spray Exposure

The salt spray exposure tests for each conductive sealant material were performed with the set of five aluminum test joints. Four of the five test joints were loaded with conductive sealant, and the other test joint (called the control joint) used no sealant. Curing of the sealant was monitored by measuring the resistance across the test joint with a double Kelvin bridge milliohmmeter, as shown in Fig. 5.7. The resistance would decrease with time, typically stabilizing after 24 hours, at which time the material was cured. The lack of melting transitions from differential scanning calorimetry (DSC) was also used to indicate total chemical crosslinking or curing.

The five (four sealant and one control) test joints were placed in an Atlas Salt Spray (Fog) Chamber, as shown in Fig. 5.8, and exposed to the environment specified in ASTM B117. The salt spray unit was operated continuously, and the resistance of the joints was measured at 24 intervals. The overall dc resistance versus weathering results are tabulated in Table 5.3. Shown in this table are the dc resistance values (mean and standard deviation) for the control samples and the sealant sample at t = 0 and t = 1000 hours. In general, it can be seen that:

1. The conductive sealant samples outperformed the control (no sealant) sample;
2. The Ag/Al and stainless steel-filled sealants outperformed the aluminum-filled sealants; and
3. The Ag/Al-filled polysilicone RTV sealant had superior dc resistance versus weathering characteristics to the other materials tested.

Fig. 5.8. Salt spray (fog) cabinet (ASTM B117) for corrosion testing of aluminum joints.

Table 5.3. Mean (X) and Standard Deviation (σ) of DC Resistance Versus Salt Spray Chamber Fog Exposure Time

Sealant Material	Manufacturer	DC Resistance (Ω)	
		0 Hours	1000 Hrs.
None (Control)	—	X = 1.0E-4	22+
Polysilicone # 1038, Ag/A[a]	Chomerics	X = 1.0E-4[f] $\sigma = 0^7$	X = 1.12E-4[f] $\sigma = 0.22$ E-4[g]
Polyurethane #4329-25-1, Ag/A[a]	Chomerics	X = 1.43E-4 $\sigma = 0.18$E-4	X = 3.63 E-2 $\sigma = 4.84$ E-2
Polyurethane #4329-25-2, Al[b,d]	Chomerics	X = 1.0E-4[f] $\sigma = 0^7$	X = 22[g] $\sigma =$ undefined[g]
Polyurethane #872, Al[b,e]	Products Research	X = 3.60E-4 $\sigma = 1.63$E-4	X = 11.18[g] $\sigma = 10.82$[g]
Proseal Polysulfide RW-2128-71, Stainless Steel[c]	Products Research	X = 2.08E-4 $\sigma = 0.53$E-4	X = 1.70E-3 $\sigma = 1.86$E-3

[a]Silver-coated aluminum powder filler
[b]Aluminum powder filler
[c]Stainless steel #316 filler
[d]Tests terminated after 360 hours due to severe corrosion
[e]Tests terminated after 336 hours due to severe corrosion
[f]Actual value lower than that indicated due to limited range of instrumentation
[g]Actual value greater than that indicated due to limited range of instrumentation

For example, the dc resistance versus salt fog exposure time for the polysilicone RTV filled with silver-coated aluminum powder" relationship is shown in Fig. 5.9. The resistance of the sealed joints stayed extremely low (on the order of 0.1 mΩ), whereas the resistance of the control joint increased measurably after 96 hours.

5.2.3 Measurement Results of Shielding Effectiveness Versus Salt Spray Exposure

Shielding effectiveness measurements were made at periodic time intervals on two of the six conductive sealant materials shown in Table 5.2: (1) the Al/Ni-filled polysulfide sealant and (2) the aluminum-filled polyurethane sealant. Shielding effectiveness was measured over the 500-1000 MHz frequency range using the test procedure outlined in Section 4.4.

The individual shielding effectiveness results were averaged over frequency and plotted as a function of salt-fog chamber exposure time. The overall results for the Al/Ni-filled polysulfide RTV sealant and the aluminum-filled polyurethane RTV sealant are shown in Figs. 5.10 and 5.11, respectively, which show the mean (de-

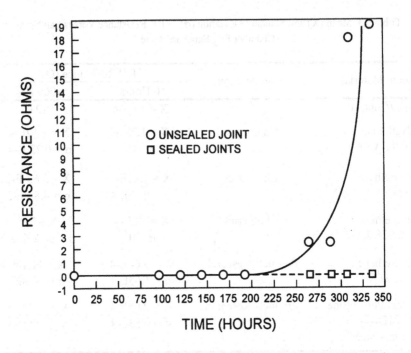

Fig. 5.9. DC resistance versus salt spray exposure time for control and Chomerics #1038 test joints.

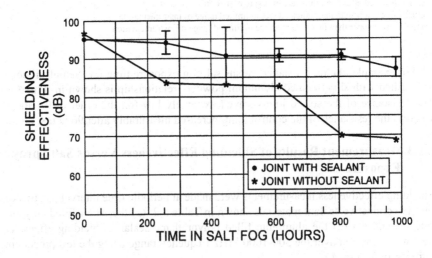

Fig. 5.10. Shielding effectiveness of control and Products Research Corp. #RW2-28-71 test joints versus salt fog exposure time.

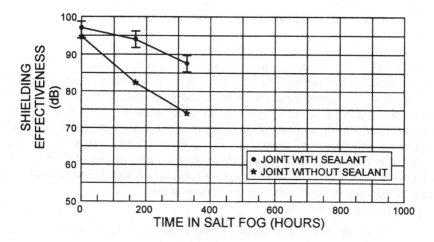

Fig. 5.11. Shielding effectiveness of control and aluminum filled polyurethane sealant test joints vs. salt fog exposure time.

noted with a dot or star) and standard deviation (denoted with a vertical line and
horizontal end caps) of the shielding effectiveness data. It can be seen from Fig.
5.10 that the shielding effectiveness of the Al/Ni-filled polysulfide material slowly
degraded from approximately 95 dB at t = 0 to approximately 87 dB after nearly
1000 hours' exposure time. The shielding performance of the stainless steel-filled
polysulfide RTV test samples were far superior to the control samples, which also
began (at t = 0) at approximately 95 dB but degraded more rapidly to approximately
68 dB after 1000 hours of exposure.

It can be seen from Fig. 5.11 that the shielding effectiveness of the aluminum-
filled polyurethane RTV sealant degraded from approximately 97 dB at t = 0 to approximately 87 dB after only 360 hours of salt fog exposure time. (In this case, the
weathering tests were terminated after 360 hours due to the presence of severe corrosion and high dc resistance readings for the test samples.) It is interesting to note
that, despite the presence of severe corrosion products and high dc resistance values,
the shielding effectiveness of the sealant samples degraded only 10 dB. Also note
that the control samples had a mean value of shielding effectiveness equal to approximately 68 dB after 1000 hours, despite the very high resistance values and
severe corrosion that took place in these joints. The primary reason for these surpris-
ingly high shielding effectiveness values is believed to be the small separation distance between the plates. For example, extrapolating the curve in Fig. 4.9 to smaller
plate separation distances, the shielding effectiveness of 68 dB, which was obtained
for the control samples after 1000 hours, could be obtained with zero dc conductivity
(i.e., an open circuit between the plates) provided that the plate separation distance
were on the order of 5 microns. It should be noted, however, that poorer shielding
performance could be expected at microwave frequencies unless uniform electrical
continuity is maintained across the entire joint to achieve low RF bond impedance.

5.3 Summary of Optimum Conductive Sealant Materials

The objective of this task was to identify and evaluate optimum corrosion prevention materials for use on nuclear hardened aircraft and weapon systems. New test joints were designed and fabricated out of 7075 aluminum and fastened using realistic torque values. The aluminum test joints were loaded with five different sealant materials and weathered in a salt spray environment according to ASTM B117. The dc resistance and shielding effectiveness of selected test joints were measured before weathering tests began and periodically during the weathering tests to monitor the corrosion effects on the electrical and electromagnetic performance of the bonds. In addition, control joints (no sealant) were weathered and tested simultaneously to compare the relative merit of the selected sealant material.

In general, it was found that:

1. the conductive sealant samples outperformed the control (no sealant) samples,
2. the Ag/Al and Al/Ni-filled sealants outperformed the aluminum-filled sealants, and
3. the Ag/Al-filled polysilicone RTV sealant had superior weathering characteristics to the other materials tests.

Based on these data, it is recommended that aluminum filler be avoided and that either silver-coated aluminum or aluminum/nickel fillers be used with an appropriate matrix (e.g., polysilicone RTV or polysulfide) for applications with aluminum substrates. The use of appropriate conductive sealant materials will improve the long-term effectiveness (both structural and electrical) of metal-to-metal bonds typically found on exposed surfaces of military aircraft and weapon systems.

5.4 Relationship Between DC Resistance, Shielding Effectiveness, and Transfer Impedance

5.4.1   Evaluation of Laboratory Sealant Test Samples

Three different materials were evaluated: (1) Chomerics 4375-27-4, (2) PRC 1764 A-2, and (3) PRC 1764 Class B. Four samples of each material were evaluated for a total of twelve different test samples. Iridited aluminum substrates were used in three of the four samples whereas one of the four samples was assembled with bare aluminum substrates. The test samples were coded for identification purposes as listed in Table 5.4. The dc resistance, shielding effectiveness, and transfer impedance of each test sample were measured. The shielding effectiveness and transfer impedance plots are contained in Appendix C and Appendix D, respectively.

Table 5.4. Test Joints and Identification.

Test joint No.	Sample No.	Joint surface coating	Sealant/part identification
1	A1	Iridite	Chomerics 4375-27-4
2	A2	Iridite	Chomerics 4375-27-4
3	A3	Iridite	Chomerics 4375-27-4
4	A4	None	Chomerics 4375-27-4
5	C1	Iridite	PRC 1764 A-2
6	C2	Iridite	PRC 1764 A-2
7	C3	Iridite	PRC 1764 A-2
8	C4	None	PRC 1764 A-2
9	D1	Iridite	PRC 1764 Class B
10	D2	Iridite	PRC 1764 Class B
11	D3	Iridite	PRC 1764 Class B
12	D4	None	PRC 1764 Class B
13	B1	Iridite	Control (No sealant)
14	B2	None	Control (No sealant)
15	B3	Iridite	Chomerics 4375-27-4

The electrical measurements were made shortly after the test samples were as-sembled and the sealants were cured. Next, the samples were placed in a salt-fog chamber for accelerated corrosion testing. The samples were periodically removed from the salt-fog chamber (ASTM B 117), dried, and the surfaces that contact the electrical test fixtures were cleaned. The dc resistance measurements were performed at approximately one-week intervals. The transfer impedance and shielding effec-tiveness measurements were repeated at approximately two-week intervals with a total salt fog exposure time of 2000 hours. A brief description of the dc resistance, shielding effectiveness, and transfer impedance measurement techniques are briefly described below.

1. DC Resistance measurements. The dc resistance of the test samples was measured with a double Kelvin bridge milliohmmeter. The dc resistance measurement procedure was described in Section 4.3. The terminals of the milliohmmeter were clipped or contacted directly to the metal plates of the test joint. All resistance measurements were made after the sealant material had cured so that the resistance value had time to stabilize.

2. Shielding effectiveness measurements. The shielding effectiveness of each test sample was determined as the difference (in dB) between the electro-

magnetic field coupled through an aperture with and without the sample present over the aperture. The shielding effectiveness measurement procedure was described in Section 4.4.

3. Transfer impedance measurements. One measure of the RF performance of shielding materials (sealants, EMI gaskets, etc.) is surface transfer impedance. Electromagnetic fields incident on the exterior of a shield induce surface currents. When these surface currents encounter an imperfectly sealed aperture or seam, a voltage is induced across the seam on the interior of the shield. These interior voltages then act as secondary sources and reradiate internal to the shield thereby degrading the performance of the shield. The transfer impedance of the conductive sealant joint may be defined as follows:

$$Z_t = \frac{V_o}{J_s} \qquad 5.1$$

where Z_t = transfer impedance (in $\Omega \cdot m$); V_o = induced output voltage (in V); and J_s = surface current density (in A/m).

The transfer impedance of each test sample was measured using a transfer impedance test fixture. The drawings for the test fixture, which has an upper frequency limited of 200 MHz, are provided in Appendix E. A block diagram of the overall transfer impedance test setup is shown in Fig. 5.12. The transfer impedance was measured over the 100-kHz to 200-MHz frequency range with an HP-3577A network analyzer. The RF output of the network analyzer was terminated in a matched 50 Ω-load resistor. The Reference Channel of the network analyzer was used to measure the input voltage that was then used to calculate the current flowing though the 50-Ω resistor and, this, the test sample. Channel A of the network analyzer was used to measure the voltage induced across the test ample. The transfer impedance was then found by dividing the voltage induced across the test sample by the current flowing though the test sample. All mathematical operations were performed by the network analyzer. The transfer impedance as a function of frequency was plotted on a digital plotter. An IBM compatible PC computer was used to control all test instrumentation.

The transfer impedance data is presented in units of dBΩ. In order to compare the data with other data measured using test fixtures with different dimensions, one must convert this data to units of $\Omega \cdot m$. Since the perimeter of these joints was approximately 53 cm, it was necessary to subtract 5.5 dB (20 log 0.530) from the data presented in the plots to yield the normalized transfer impedance.

4. Relationship of transfer impedance to shielding effectiveness. Shielding effectiveness is dependent on many variables including the properties of the shield (shield geometry, seam dimensions, seam orientation) as well as the

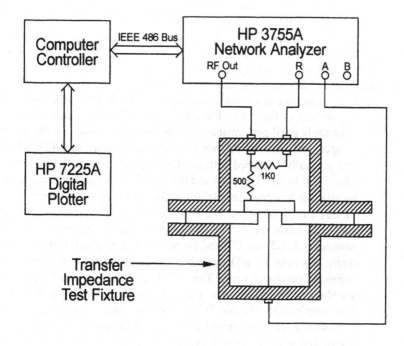

Fig. 5.12. Block diagram of the transfer impedance test setup.

properties of the electromagnetic wave (impedance, polarization, and inci-
dence angle). Therefore, the relationship between Z_t and SE can be extremely complex for practical shield geometries and field conditions. However, as a simple rule of thumb, the following relationship was used:

$$SE(dB) = 22.5 - Z_t(dB\Omega \cdot m)$$ 5.2

The constant (22.5 dB$\Omega \cdot$m) in Eq. 5.2 was empirically derived from
the measured data and thus applies to the particular geometry and field conditions used during our tests. Since the upper frequency limit of the transfer
impedance measurements was 200 MHz and the lower frequency limit of
the shielding effectiveness measurements was 500 MHz, the 200-MHz
transfer impedance data was extrapolated to 500 MHz using the 200-MHz
amplitude and slope. The constant in Eq. 5.2 was then calculated each test
sample by adding the 500-MHz shielding effectiveness value to the extrapolated 500-MHz transfer impedance value. The resultant values were then
statistically analyzed to have an average value of 22.5 dB$\Omega \cdot$m and a standard deviation of 3.0 dB. As an example, Sample A2 (t = 377 hrs) had a
transfer impedance of -52.5 dBΩ at 200 MHz and an extrapolated transfer
impedance of -50 dBΩ at 500 MHz. Normalizing this extrapolated imped-

ance by the perimeter length yielded a transfer impedance of -55.5 dB$\Omega \cdot$ m which, when inserted in Eq. 5.2, yielded a predicted shielding effectiveness at 500 MHz of 78 dB. This predicted value was close to the measured shielding effectiveness value for Sample A2 at 500 MHz of 75 dB.

5. Summary of laboratory test results.

   (a) D. C. resistance test results. A summary of the de resistance test results is provided in Table 5.5. The Chomerics 4375-27-4 samples provided reasonably good performance. The dc resistance values for the iridited samples (A1, A2, and A3) were initially in the range of 0.42–0.67 mΩ and gradually increased with salt-fog exposure time to 1.17–2.50 mΩ after 2000 hours. The none-iridited sample (A-4) initially had an immeasurable low resistance (< 0.1 mΩ), but the resistance increased to 28 mΩ after 2000 hours. The PRC 1764 A-2 sealant samples provided excellent performance. The dc resistance values for both the iridited samples (C1, C2, and C3) and the non-iridited ample (C4) were consistently less than 1.5 mΩ through the duration of the testing and were typically less than 0.1 mΩ. The PRC 1764-B sealant samples also provided excellent performance. The dc resistance values for both the iridited samples (D1, D2, and D3) and the non-iridited sample (D4) were consistently less than or equal to 1.3 mΩ though the duration of the testing and were often times less than 0.1 mΩ.

   (b) Shielding effectiveness test results. A summary of the shielding effectiveness test results is provided in Table 5.6. The shielding data was averaged over the 500–1000 MHz frequency range and this average shielding value was used in the summary table. The Chomerics 4375-27-4 samples provided reasonably good shielding performance over the 2000 hours salt-fog test period. The average shielding effectiveness values for the iridited samples (A1, A2, and A3) were initially in the range of 76 to 88 dB and gradually decreased with alt-fog exposure time to 61–78 dB at 2000 hours. The non-iridited sample (A-4) initially had a immeasurably high shielding effectiveness value (> 90 dB) but the shielding decreased rapidly with salt-fog exposure to a value of 55 dB after 2000 hours. The PRC 1764 A-2 sealant samples provided excellent shielding performance. The shielding effectiveness values for both the iridited samples (C1, C2, and C3) and the non-iridited sample (C4) exceeded the measurement dynamic range (> 90 dB) through the duration of the testing. The PRC 1764-B sealant samples also provided excellent shielding performance. The shielding effectiveness values for both the iridited samples (D1, D2, and D3) and the non-iridited sample (D4) were consistently greater than 90 dB for the duration of the test period.

Table 5.5. DC Resistance as a function of salt fog exposure time.

Sealant	Sample	DC Resistance (mΩ)											
		Hours of Salt Spray Exposure											
		0	164	337	517	681	800	960	1124	1264	1428	1572	2000
Chomerics 4375-27-4	A1	0.67	0.71	0.92	0.80	0.89	0.93	0.97	0.98	1.06	1.10	1.20	11.59
	A2	0.47	0.80	0.73	0.52	0.69	0.77	0.96	1.02	1.12	1.12	1.46	2.5
	A3	0.42	0.70	0.70	1.15	0.62	0.66	0.74	0.71	0.77	0.80	0.86	1.17
	A4	<0.1	0.23	0.22	<0.1	<0.1	<0.1	0.10	0.10	0.17	0.21	0.29	28
PRC 1764A-2	C1	0.11	1.10	0.30	<0.1	<0.1	<0.1	<0.1	<0.1	<0.1	<0.1	<0.1	<0.1
	C2	<0.1	1.47	0.35	<0.1	<0.1	<0.1	<0.1	<0.1	<0.1	<0.1	<0.1	<0.1
	C3	0.14	0.24	0.37	<0.1	<0.1	<0.1	<0.1	<0.1	<0.1	<0.1	<0.1	<0.1
	C4	<0.1	0.18	0.37	<0.1	<0.1	<0.1	<0.1	<0.1	<0.1	<0.1	<0.1	<0.1
PRC 1764 Class B	D1	0.36	1.30	0.20	<0.1	<0.1	<0.1	<0.1	<0.1	<0.1	<0.1	<0.1	<0.1
	D2	0.17	0.43	0.18	<0.1	<0.1	<0.1	<0.1	0.12	0.12	0.17	0.18	0.1
	D3	<0.1	0.45	0.14	<0.1	<0.1	<0.1	<0.1	<0.1	<0.1	<0.1	<0.1	<0.1
	D4	<0.1	0.15	<0.1	<0.1	<0.1	<0.1	<0.1	<0.1	<0.1	<0.1	<0.1	<0.1
					468 hr.	612 hr.	1285 hr.						
None	B1	0.15	0.15	0.18	<0.1	0.23	0.31						
None	B2	<0.1	<0.1	0.21	0.24	0.68	>22,000						
Chomerics 4375-27-4	B3	0.38	0.40	0.41	0.42	0.45	0.50						

Non-iridited surfaces: A4, C4, D4, and B2.

Table 5.6 Average shielding effectiveness (dB) of test samples.

Sealant	Sample Number	Hours of salt spray exposure				
		681	960	1264	1572	2000
Chomerics	A1	75	75	75	73	71
1375-27-4	A2	70	67	67	66	61
	A3	79	77	77	75	73
	A4	90 ^{L}	82 ^{L}	69	63	55
PRC	C1	90 ^{L}	90 ^{L}	90 ^{L}	90 ^{L}	90 ^{L}
1764 A-2	C2	90 ^{L}	90 ^{L}	90 ^{L}	90 ^{L}	90 ^{L}
	C3	90 ^{L}	90 ^{L}	90 ^{L}	90 ^{L}	90 ^{L}
	C4	90 ^{L}	90 ^{L}	90 ^{L}	90 ^{L}	90 ^{L}
PRC	D1	90 ^{L}	90 ^{L}	.90 ^{L}	90 ^{L}	90 ^{L}
1764-B	D2	90 ^{L}	90 ^{L}	90 ^{L}	90 ^{L}	90 ^{L}
	D3	90 ^{L}	90 ^{L}	90 ^{L}	90 ^{L}	90 ^{L}
	D4	90 ^{L}	90 ^{L}	90 ^{L}	90 ^{L}	90 ^{L}
		612 hr.		1285 hr.		
None	B1	85		90		
None	B2	63		72		
Chomerics 4375-27-4	B3	75				

Note: Entries with the superscript L indicate that the reading was limited by the dynamic range of the test set-up, and that the actual shielding effectiveness was probably *greater* than the reported value.

(c) Transfer impedance test results. A summary of the transfer impedance test results is provided in Table 5.7. Separate entries are provided for 100 kHz (lowest test frequency) and 200 MHz (highest test frequency). Note that the low frequency (i.e., 100 kHz) transfer impedance corresponds to the dc resistance of these test joints. The Chomerics 4375-27-4 samples provided reasonably good performance. The transfer imped-ance values for the iridited samples (A1, A2, and !3) were initially in the range of -65 to -54 dBΩ (0.6–2.0mΩ) and gradually increased with salt-fog exposure time to -59 to 36 dBΩ (1.1–15.8 mΩ) at 2000 hours. The non-iridited sample (A-4) initially had transfer impedance values of -87 (or less) to -79 dBΩ (0.04–1.1 mΩ) but the transfer impedance increased rapidly to a range from -32 to 22 dBΩ (25.1–79.4 mΩ) after 2000 hours. The PRC 1764 A-2 sealant samples provided excellent per-formance. The transfer impedance values for both the iridited samples (C1, C2, and C3) and the non-iridited sample (C4) ranged from -99 to -70 dBΩ (0.01–0.3 mΩ) and showed very little degradation in per-formance after 2000 hours. The PRC 1764-B sealant samples also pro-

vided excellent transfer impedance performance. The transfer impedance values for both the iridited samples −91 to −68 dBΩ (0.03–0.4
mΩ) and non-iridited samples showed very little degradation in
performance after 2000 hours.

Based on the results obtained from the experimental investigations,
a relaxation of the 2.5-mΩ requirement is neither justifiable nor does it
appear to be necessary. The results of the empirical study of the relationship between dc resistance and shielding effectiveness (see Fig.
4.7) indicate that relaxation of the bonding resistance requirement to 10
mΩ could result in a 20-30 dB reduction in shielding effectiveness
(relative to a 2.5 mΩ bond) and the relaxation of the bonding resistance
to 100 mΩ could result in a 40–55 dB reduction in shielding effectiveness. Furthermore, based on the empirical study of the effects of weathering on electrical performance of the bond, conductive sealants tested
from two different vendors silver-coated aluminum-filled polysilicone
RTV and Al/Ni-filled polysulfide) maintained a bonding resistance
well under the 2.5 mΩ requirement with little or no corrosive performance degradation after 1000 hours in the salt fog environment per
ASTM B117. Consequently, relaxation of the 2.5 mΩ requirement was
not recommended as part of the proposed modifications.

The interested reader is referred to the Final Reports of Georgia
Tech Research Institute Project Numbers A-4324 and A-8245, Contract
Numbers F02060-85-66007 and F09603-88-R-2808, WR-ALC, Robins
Air Force Base.

Table 5.7. Transfer impedance (dB-Ω) of test samples.

Sealant	Sample	Salt spray exposure (hr.)							
		0		377		681		960	
		100 kHz	200 MHz	100 kHz	200 MHz	100 kHz	200 MHz	100 kHz	200 MHz
						Test frequency			
Chomerics 4375-27-4	A1	-61	-54	-61	-52.5	-58	-51	-58	-51
	A2	-65	-61	-64	-44	-61	-39	-57	-36
	A3	-63	-60	-64	-56	-61	-54	-61	-54
	A4	-87[L]	-79	-78[L]	-75[L]	-78	-65	-75	-58
PRC 1764 A-2	C1	-76	-95[L]	-77[L]	-91[L]	-72	-85[L]	-72	-77[L]
	C2	-79	-89[L]	-77[L]	-79[L]	-76	-79[L]	-76	-85[L]
	C3	-79	-91[L]	-75[L]	-83[L]	-76	-86[L]	-70	-82[L]
	C4	-88[L]	-99[L]	-79[L]	-79[L]	-79	-79[L]	-73[L]	-75[L]
PRC 1764-B	D1	-71	-87[L]	-70	-83[L]	-68	-85[L]	-69	-82[L]
	D2	-72.5	-82[L]	-72	-82[L]	-71	-82[L]	-68	-72[L]
	D3	-76[L]	-81[L]	-74	-87[L]	-76	-83[L]	-73	-74[L]
	D4	-91[L]	-88[L]	-79[L]	-79[L]	-84[L]	-83[L]	-78[L]	-73[L]

Sealant	Sample	0		304		612 hr.	
		100 kHz	200 MHz	100 kHz	200 MHz	100 kHz	200 MHz
None	B1	-71	-60	-71	-60	-70	-57
None	B2	-77[L]	-68[L]	-71	-42	-57	-34
Chomerics 4375-27-4	B3	-63	-51	-68	-47	-67.5	-51

Table 5.7. Continuation.

Sealant	Sample	Salt spray exposure (hr.)					
		1264		1572		2000	
		Test frequency					
		100 kHz	200 MHz	100 kHz	200 MHz	100 kHz	200 MHz
Chomerics 4375-27-4	A1	-60	-51	-59	-51	-56	-48
	A2	-58	-36	-57	-37	-52	-36
	A3	-62	-53	-61	-53	-59	-52
	A4	-75	-43	-67	-38	-32	-22
PRC 1764 A-2	C1	-82[L]	-83[L]	-82[L]	-92[L]	-79	-87[L]
	C2	-83[L]	-80[L]	-86[L]	-96[L]	-80	-83[L]
	C3	-79[L]	-79[L]	-85[L]	-96[L]	-78	-88[L]
	C4	-84[L]	-84[L]	-78	-89[L]	-74	-90[L]
PRC 1764-B	D1	-85[L]	-78[L]	-81	-92[L]	-79	-90[L]
	D2	-86[L]	-76[L]	-79	-92[L]	-76	-100[L]
	D3	-83[L]	-83[L]	-83[L]	-96[L]	-80	-89[L]
	D4	-81[L]	-84[L]	-84[L]	-94[L]	-77	-85[L]
		1285 hr.					
None	B1	-65	-47				
None	B2	32	-17				
Chomerics 4375-27-4	B3	-66	-52				

Note: Entries with the superscript L indicate that the reading was limited by the measurement sensitivity and that the transfer imped-ance was probably *lower* (i.e., more negative) than entered value.

6

Field Test Evaluations on E-3A Aircraft

6.1 Introduction

At the present, chemical conversion coating (iridite) is used on the aluminum faying surfaces as the only protective finish that meets the EMP/EMI requirements. This procedure allows moisture to be trapped between the surfaces, resulting in severe corrosion. The nonconductive corrosion products that develop between these surfaces have proven to increase the impedance high enough to destroy the nuclear hardness of an weapon system.

A test plan for conducting field tests on active aircraft was developed. The purpose of these tests was to evaluate the effectiveness in the field of selected conductive sealant materials with respect to corrosion and EMP/EMI requirements. First, several bare aluminum faying surface on each system were selected for application of the conductive sealant materials. Nest, the resistance of each joint was measured with a double Kelvin bridge milliohmmeter before application of the sealant material. The conductive sealant material was then liquid-applied on the bare aluminum faying surfaces, cured, and the resistance of the joint was remeasured. The resistance of the joints was measured again after a one-year period.

Corrosion effects were determined by visual inspection and by field optical microscope with photographic capability. EMP/EMI effectiveness evaluation was not nearly as straightforward as corrosion evaluation. DC resistance measurements are the most practical measurements that can be made for an actual installation (which is the primary reason why resistance is the parameter specified in MIL-B-5077B for evaluation of bond effectiveness by the contractor). As described above, dc resistance measurements were performed before and after application of the sealant materials and after a two-year period. Complicating the dc resistance measurement was the problem that the actual resistance of the joint containing the conductive sealant materials was shorted out by either the conductive fasteners used to construct the joint or by a variety of other parallel paths set up by metallic structures which are in contact with both faying surfaces.

It should be noted (as described in the Georgia Tech Research Institute final technical report[m] on Contract No. F09603-85-R-0959) that the shielding effectiveness of a joint, and hence the effectiveness against EMP/EMI sources, cannot be determined simply from a knowledge of the de resistance. A low dc resistance is not sufficient for good shielding effectiveness. Electrical continuity must be obtained across the entire joint area for effective shielding over a broad frequency range. This requirement is not necessarily met by a low dc resistance value nor was it to be verified solely by a dc resistance measurement. Consequently, measurements in addition to the required dc resistance measurements were beneficial in terms of evaluating the effectiveness of the conductive sealants with respect to EMP/EMI requirements.

Additional means of characterizing the bonding effectiveness of conductive sealants included shielding effectiveness measurements and transfer impedance measurements. Shielding effectiveness testing is the most direct method of evaluating conductive sealants performance with respect to EMP/EMI requirements. However, in certain instances, shielding effectiveness testing may not be feasible. For example, adequate room may not exist inside the missile to place the receiving antenna that is required to perform standard (e.g. MIL-STD-285) shielding effectiveness measurements. Another difficulty surrounds the fact that illumination of the system during shielding effectiveness measurements will expose other shield flaws, which will make isolation of the joint-under-test extremely difficult or impossible. Consequently, the use of shielding effectiveness measurements was carefully assessed in terms of its feasibility and applicability to bonding effectiveness evaluation of the conductive sealant materials.

An approach, which was applicable to the present problem, is to measure the transfer impedance of the conductive sealant bonds over a wide frequency range. Transfer impedance measurements made over a broad frequency range and at several different points along the length of the joint would provide a better indication of bond effectiveness than a dc resistance measurement. Relative comparisons were made between bonds with conductive sealants applied versus those without conductive sealants applied.

The authors, representing GTRI, proposed to perform the required dc resistance measurements and to supplement these measurements with additional measurements that more effectively evaluated the effectiveness of the sealant materials in protecting against EMP/EMI sources. Selection of the particular joints to be tested included consideration as to the testability of the joint from both a dc resistance and an transfer impedance/shielding effectiveness measurement perspective. The existence of parallel conducting paths were identified and documented, and their impact on the dc resistance measurements was assessed.

6.1.2 Visits to Air Force Installations

The authors visited the Oklahoma Air Logistics Center in order to conduct field testing on identified aircraft. The first visit to this installation was made early in the program. During this first visit, the panels/joints were selected and all of the necessary information regarding these panels (such as location, total faying surface area, fastening technique, number of fasteners, existence of parallel conducting paths, etc.) was obtained for test planning purposes. After the test plans were complete, a second visit was made to apply the sealant material and to make the electrical performance measurements both before and after application of the sealants. The third visit was planned to be made one year after the second visit to re-evaluate the performance of the conductive sealant materials. However, the final inspection was not possible for 24 months since the plane was on mission.

The areas that were selected on the E-3A aircraft for testing of the conductive sealants are listed below and are shown in relationship to the aircraft in Figs. 6.1 and 6.2:

1. Vertical Tail Fin (see Fig. 6.3 and 6.4). Corrosion in the form of pitting around the fastening surface of the fin has been regularly observed.
2. Forward Entry Door (see Fig. 6.5). Severe pitting adjacent to the elastomeric gasket has been regularly observed, especially on the bottom sections where condensation collects.
3. Aft Entry Door (see Fig. 6.6). Same as in b.
4. Rotodome Air Inlet Duct EMP Shield (see Fig. 6.7). The vent has been observed to corrode badly and produce larger passages due to failure of the metal.

All of these areas have exhibited well-documented and notorious corrosion problems. The bonds in these areas all required electrical conductivity. The authors theorized that, if the proposed conductive sealants could solve the corrosion problem, then the electrical conductivity and shielding effectiveness could be maintained as well.

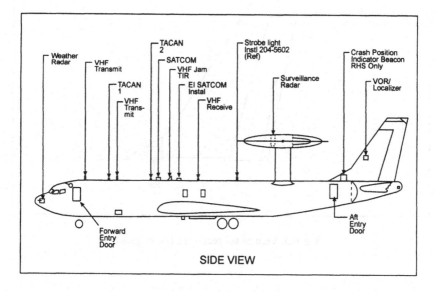

Fig. 6.1. Profile of E-3A aircraft.

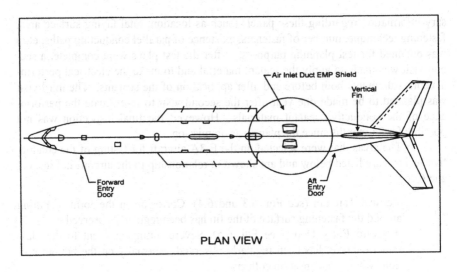

Fig. 6.2. Top view of E-3A aircraft.

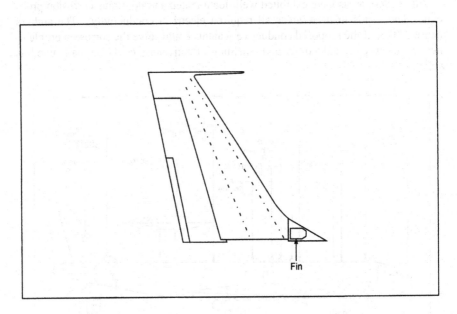

Fig. 6.3. Vertical tail section and patch panel.

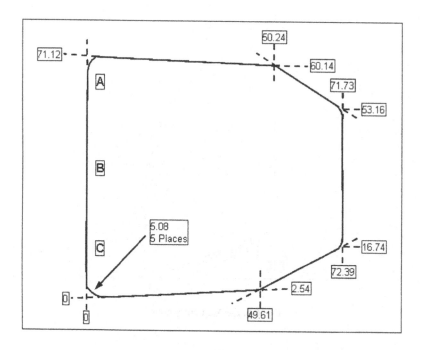

Fig. 6.4. Patch panel dimensions (ISO dimensions, cm).

Fig. 6.5. Front-left view of E-3 aircraft showing forward and aft entry doors.

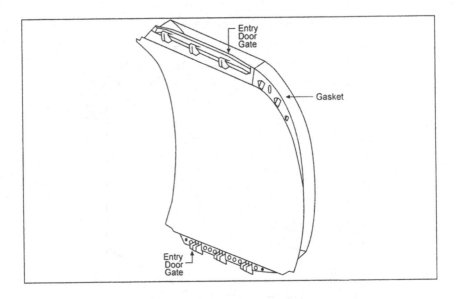

Fig. 6.6. E-3A aft entry door.

Fig. 6.7. Underside of rotodome. Air inlet duct emp shield (highlighted by white dashed lines) located inside the dome and accessible through entrance hatch.

Table 6.1. Location of test sites on E-3A aircraft and conductive sealants applied.

Test Site	Conductive Sealant
Entry door, forward	Products research 1764 A/B
Entry door, aft	Products research 1764 A/B
Air vent EMP shield, rotodome	Chomerics 4375-27-4 A/B
Vertical tail fin	Chomerics 4375-27-4 A/B

6.2 Field Test Description

6.2.1 Location of Test Sites

The location of test sites on the E-3A aircraft with corresponding conductive sealants are listed in Table 6.1.

The test sites on the E-3A aircraft are shown in following figures:

1. Figure 6.8: Front left view of E-3A aircraft showing forward and aft entry doors.
2. Figure 6.9: View of the bottom section of the forward entry door, gasket and corrosion area.
3. Figure 6.10: Installation of probes from measurement of dc resistance on bottom section of forward entry door.
4. Figure 6.11. Forward - right view of the E-3A aircraft (arrow showing vertical tail fin panel).
5. Figure 6.12: View of the tail section of the E-3A aircraft, right side, showing vertical tail fin panel and corrosion area.
6. Figure 6.13: View of the tail section, left side, showing access panels for measuring transfer impedance.
7. Figure 6.14: Preparation of vertical tail fin panel (removal of paint to bare metal) for attachment of transfer impedance fixture.
8. Figure 6.15: Installation of transfer impedance fixture.
9. Figure 6.16: View of fixture and instruments for measuring transfer impedance.

6.2.2 Field Test Methods

DC resistance and transfer impedance measurements were performed on the E-3A aircraft. In order to perform the transfer impedance measurements in the field, a specialized field transfer impedance test fixture was designed and fabricated. A photograph of the field test fixture is shown in Fig. 5.6. In order to validate the functionality of the field test fixture, transfer impedance measurements on laboratory samples were performed with both the laboratory and field test fixtures. Comparative plots of the lab and field test fixtures for high impedance ($R_{dc} = 1\ \Omega$) and low impedance ($R_{dc} = 5$ mΩ) test samples are shown in Figs. 6.17 and 6.18, respectively. Note that the close agreement occurs at low frequencies where the "effective" perimeter is essentially the same for both fixtures. However, at high frequencies, the

Fig. 6.8. Front left view of E-3A aircraft showing forward and aft entry doors.

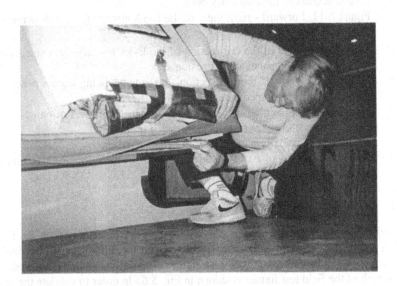

Fig. 6.9. View of the bottom section of the forward entry door, gasket and corrosion area.

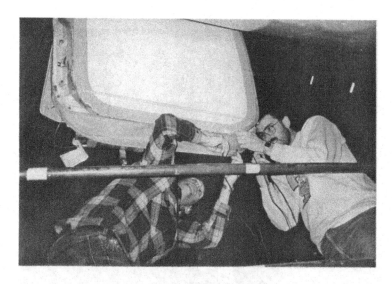

Fig. 6.10. Installation of probes for measurement of dc resistance on bottom section of forward entry door.

Fig. 6.11. Forward–right view of the E-3A aircraft (arrow showing vertical tail fin panel).

Fig. 6.12. View of the tail section of the E-3A aircraft, right side, showing vertical tail fin panel and corrosion area.

Fig. 6.13. View of tail section, left side, showing access panels for measuring transfer impedance.

Fig. 6.14. Preparation of vertical tail fin panel (removal of paint to bare metal) for attachment of transfer impedance fixture.

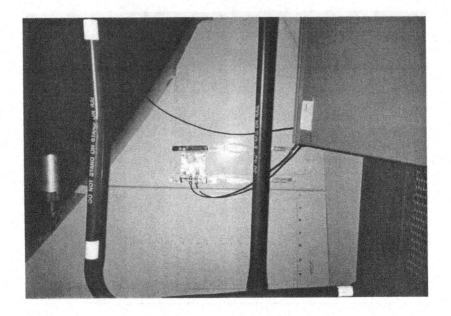

Fig. 6.15. Installation of transfer impedance fixture.

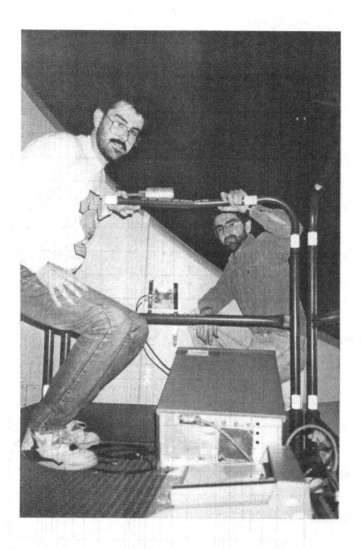

Fig. 6.16. View of fixture and instruments for measuring transfer impedance.

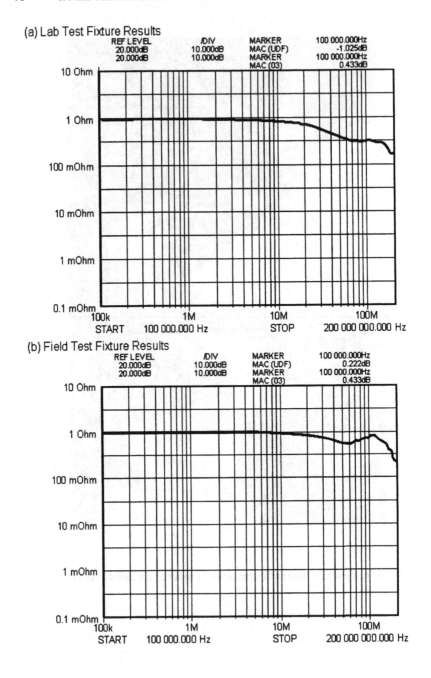

Fig. 6.17. Comparison of measured Z_t using laboratory and field test fixtures (R_{dc} = 1 ohm).

(a) Lab Test Fixture Results

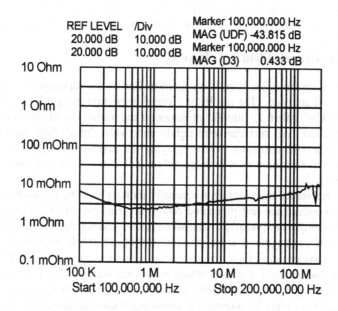

(b) Field Test Fixture Results

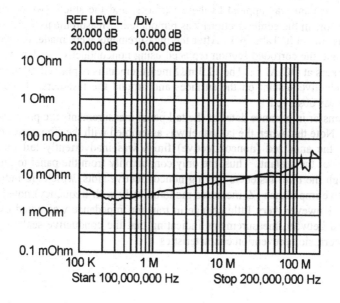

Fig. 6.18. Comparison of measured Z_t using laboratory and field test fixtures (R_{dc} = 5 milliohms).

field test fixture essentially concentrates surface current over a smaller region resulting in an increase in impedance of approximately 10 dB (implying a reduction in effective perimeter from the lab fixture value of 0.53 m to a value of 0.17 m). Overall, the field test fixture provides reasonable and accurate transfer impedance data.

6.2.3 E-3A Aircraft Field Test Results

The test areas on the E-3A aircraft are listed in Table 6.1 with respect to the type of conductive sealant used. The results of a visual inspection using a field microscope are contained in Table 6.2. Significant corrosion and pitting was observed during the initial inspection. All surfaces were lightly abrasively cleaned with sand paper to remove products of corrosion before applying the conductive sealants. After 24 months, the same surfaces were re-inspected no indications of corrosion were present in all cases.

The dc resistance measurements generally indicate that the conductive sealants maintained conductance and prevented corrosion. However, there are some measurements that require explanation. The aft entry door did not show corrosion after 24 months of service in contrast to the initial inspection, but the dc resistance rose to 4.6 $m\Omega$ in one measurement. First, the dc resistance measurements are actually a measurement across the metal fasteners where no sealant was used; and a measurement across the sealant and fasteners after 24 months. Severe corrosion was observed initially and none after 24 months. Considering these circumstances, the authors believes that the 1.4 - 4.6 $m\Omega$ measurements is an indication of good conductivity and corrosion prevention although 2.5 $m\Omega$ and below would be preferred.

Referring to the air vent EMP shield, the corroded surfaces were cleaned, the conductive sealant was applied to these surfaces, and the shield (which had some severe damage in the center section) was remounted. Measurements were made and reported as shown in Table 6.3. After the measurements were made, the shield was removed, and the corroded section replaced using nonconductive sealant by maintenance workers at OC-ALC. The 11.3 $m\Omega$ measurement is explained by the presence of nonconductive sealant on the surfaces and around the fasteners. No additional corrosion was observed.

The transfer impedance data for the tail panel measurements are provided in Appendix F. Note that when the tail panel was assembled with the Chomerics 4375-27-4 sealant, the anodized (nonconductive) finish was inadvertently left on prior to application of the sealant. Thus, the only conductivity from the panel to the aircraft was through the rivet fasteners leaving nonconductive slots, which are indicated by the inductive impedance (Z_t increases monotonically with frequency) noted in all of the plots. However, very little change in transfer impedance was noted over the 2 year period between measurements indicating that the conductive sealant provided excellent corrosion prevention characteristics.

Table 6.2. Corrosion areas on E-3A aircraft and results of visual inspections.

Area on E-3	Results of visual inspections	
	Before treatment October 28, 1989	After treatment October 12, 1991
Entry Door, Forward	0.12 in. pits along surfaces	No additional pitting
Entry Door, Aft	0.12 in. pits along surfaces	No additional pitting
Air vent EMP shield	Severe corrosion in shield[1]	No additional pitting
In rotodome	(vent was partially destroyed)	
Vertical tail pin	0.03 in. pitting observed	No additional pitting

6.3 Conclusions

These conclusions are based on the previous laboratory study, Electromagnetic Pulse and Electromagnetic Interference Versus Corrosion Protection Methods (31 August 1987), the above field report. The above field study utilized the successful conductive sealant materials for application on an E-3 aircraft. Visual inspection, and electrical measurements including dc resistance and transfer impedance, were performed on four corrosion prone areas to measure the effectiveness of the sealant materials.

The conclusions follow:

Table 6.3. Results of DC resistance and rransfer impedance measurements on E-3A aircraft.

Area on E-3	DC resistance (mΩ)	Transfer impedance (mΩ)	DC resistance (mΩ)	Transfer impedance (mΩ)
	Untreated/ treated	Untreated/ treated	24 months	24 months
Entry door, forward (product research 1764)	0.62–11.1/ 0.23–0.37	Not tested	0.49–0.62	Not tested
Entry door, aft (product research 1764)	1.02–1.28/ 0.48–0.50	Not tested	1.4–4.6	Not Tested
Air vent EMP shield in rotodome (Chromerics 4375-27-4)	1.45–5.50/ 0.10–0.10	Not tested	1.4–11.3[1]	Not Tested
Vertical tail fin (Chomerics 4375-27-4)	0.40–0.40/ 0.39–0.39	(A) 3.20–1000/ 0.32–56.00 (B) 1.80–0.32/ 0.40–56.00 (C) 2.50–0.45/ 0.10–1000)	0.20–1.9	(A) 4.63–89 (B) 0.16–119 (C) 0.16–1000

1. A visual corrosion was arrested on all areas tested using both conductive sealant materials.
2. Transfer impedance and dc resistance measurements indicated that the corrosion was arrested in all test areas.
3. Corrosion was successfully monitored using dc resistance and transfer impedance; a correlation was made.
4. The 2.5 mΩ dc resistance requirement is achievable using conductive sealants and for up to at least 24 months of field service.
5. Salt Spray (ASTM B117) testing on test panels correlated to field results; the accelerated method is an acceptable method of evaluating materials for this application.
6. The use of the proven conductive sealants have the potential of saving the Air Force significant maintenance time in replacing parts and costs of parts/materials.

6.4 Recommendations

Based on the successful results of the above report and those reported in reference ???., we recommend the following:

1. Use of the tested sealant materials on corrosion - prone joints and bonds which require the dc 2.5 mΩ dc resistance. The authors feels comfortable in recommending these materials for extensive use on equipment which will result in significant saving in parts replacement maintenance.
   Applicable equipment could include:

   (a) aircraft
   (b) missiles
   (c) electronic/communications equipment
   (d) ground equipment

2. Pretesting and evaluation of sealant materials using the GTRI test joints and methods.
3. An advanced field study using conductive sealants extensively on test aircraft to confirm the cost saving advantages of corrosion inhibiting and conductive polymeric sealant.
4. A specification (see Appendix G) that is refined and developed for acquisition use; and an American Materials Specification (AMS) developed from this specification.

Assessment of the Validity of the MIL-B-50878 Class R Bonding Requirements

7.1 Background

The electrical bonding requirements for limiting RF potentials (Class R bonding) as set forth in MIL-B-5087B, "Military Specification -- Bonding, Electrical and Lightning Protection for Aerospace Systems," required that the vehicle skin be designed to produce a uniform low-impedance surface through inherent RF bonding and that the contractor demonstrate by test that the bonding method results in a dc resistance of less than 2.5 mΩ. In addition, current Air Force design practice (per AFSC Design Handbook DH 1-4, Design Note 5D1) required the contractor to meet the conductivity requirements independent of the use of conductive fasteners, thereby necessitating intimate metal-to-metal contact. Consequently, the finishes which are commonly used to meet existing electromagnetic pulse (EMP)/electromagnetic interference (EMI) requirements for aluminum surfaces are tin plating or MIL-C-5541, Type 3 chemical conversion coatings. However, these methods allow moisture to be trapped between metal surfaces, which can lead to severe corrosion problems in the field. -- Corrosion between metal surfaces has reportedly been so severe as to create structural weaknesses which undermine the effectiveness of the structure[1]. Further, the products of corrosion are nonconductive materials, which increase the electrical resistance/impedance of the bond and, thus, can destroy the nuclear hardness of aircraft and weapon systems.

Consequently, the 2.5-mΩ bonding requirement was questioned. If a higher resistance value could provide the necessary conductivity for nuclear hardening, a variety of conductive sealant materials would be available which could be used to reduce corrosion between faying surfaces. Furthermore, the use of conductive sealants could enable a low resistance and effective bond (both electrically and mechanically) to be maintained over the useful life of the weapon system. Therefore, a study was undertaken to:

1. determine the basis and rationale behind the 2.5 mΩ bonding requirement,
2. determine the relationship (if any) between the dc resistance and EMI/EMP hardness of a bonded joint, and
3. perform measurements to determine the effects of weathering on the dc resistance and EMI/EMP hardness of aluminum joints bonded with and without the use of conductive sealant materials.

7.2 Basis/Rational Behind the 2.5-Milliohm Bonding Requirements

The basis and rational behind the 2.5 mΩ requirement given in MIL-B-5087B for Class R bonding was investigated. A thorough review of bonding-related documents was undertaken, and valuable information was also obtained from personal communications with individuals involved with the development of the original MIL-B-5087 bonding specification. The most useful information came from discussion with Mr. C. E. Seth of Wright Patterson Air Force base, Aeronautical Systems Division (WPAFB/ASD). WPAFB/ASD was the custodian of the original MIL-B-5087 specification, and Mr. Seth was involved in the development of this specification from its inception. Much of the information with regard to the original purpose and intent of the 2.5-mΩ requirement was obtained from Mr. Seth of WPAFB/ASD.

The original purpose of the 2.5-mΩ requirement was to ensure adequate bonding of metal joints to provide lightning protection, RF shielding and system electromagnetic compatibility (EMC). The rational behind this requirement was that the 2.5 mΩ dc resistance was a value that:

1. could be readily achieved using accepted bonding practices,
2. could be readily verified through measurement, and
3. would force the contractor to properly clean the surfaces before assembly and maintain adequate pressure between the mating surfaces to ensure good metal-to-metal contact.

Higher values of resistance tend to relax the bond preparation and assembly requirements.

One of the oldest cited references to the 2.5 mΩ resistance limit is Volume I of a report prepared in 1964 by the Filtron Company, Inc., for the U. S. Army Electronics Laboratories. This report, which is entitled, "Interference Reduction Guide for Design Engineers," states on pages 2-20 that,

"The dc resistance of an adequate bond should be between 0.00025 and 0.0025 ohm."

No further rationale is provided for the recommended range of dc resistances. However, the authors do concede that

"The effectiveness of a bond at radio frequencies is neither fully dependent upon nor measurable only in terms of its dc electrical resistance; especially at high frequencies..."

and that dc measurements are employed since

"...it is more convenient to measure the dc resistance rather than the ac impedance of a bond...".

The overall role and function of bonding is discussed in Design Note 5D1 of AFSC DH 1-4:

"Electrical bonding is the process of mechanically connecting certain metal parts so that they will make a good low-resistance electrical contact. Bonding is required to ensure that a system is electrically stable and relatively free from the hazards of lightning, static discharge, and electrical shock and to assist in the suppression of RF interference. Usually, the resistance of electrical bonds should be in the order of 0.0025 ohm."

The implication that a 2.5-mΩ dc resistance is necessary for adequate bonding is cited over and over again in the literature. However, no technical basis is given for this particular resistance value.

The 2.5-mΩ bonding requirement specified in MIL-B-5087B pertains to Class R bonds whose function is to limit RF potentials. Different dc resistance values are specified for other bond classifications depending on their primary function. These requirements are summarized in MIL-HDBK-253, "Guidance for the Design and Test of Systems Protected Against the Effects of Electromagnetic Energy," on page 15, which states:

"Measurement of the de resistance of a bond is often used as a guide to the anticipated performance of the bond. Depending on the purpose of the bond, some military documents specify the maximum dc resistance allowable for a good bond. For example, bonds that are installed to prevent shock hazards are required by both MIL-B-5087 and MIL-STD-1310 to have a resistance of less than 0.1 ohm. Bonds for RF purposes are required by MIL-B-5087 to have a resistance of less than 2.5 milliohm. Additionally, in areas prone to explosion or fire hazards, maximum values of bond resistances are designated; these values are a function of anticipated maximum fault current in the event a power line to ground short occurs. A guideline as far as a good RF bond is concerned is a dc resistance value of between 0.25 and 2.5 milliohm."

Again, the same range of dc resistance values (0.25 to 2.5 mΩ), as in the Filtron report, are recommended for adequate bonding.

Several other documents, in addition to MIL-HDBK-253, AFSC DH 1-4 and the Filtron report, also specify the 2.5 mΩ bonding resistance or reference MIL-B-5087B. These documents include North Atlantic Treaty Organization (NATO) Standardization Agreement No. 3659, MIL-STD-462, MIL-STD-1310, MIL-STD-1512, MIL-STD-1541, MIL-STD-4544 and MIL-E-6051D.

In summary, a large number of documents have been reviewed which appear to substantiate the information obtained from the personal communications. Based on these sources of information, it is concluded that the 2.5-mΩ bonding requirement arose, not from a theoretically or empirically derived basis, but from a practical standpoint and as a means for ensuring proper bond preparation and assembly.

7.3 Summary and Conclusions on Validity of 2.5 Milliohm Bonding Requirements

In summary, the basis and rationale behind the 2.5-mΩ bonding requirement in MIL-B-5087 has been determined from an extensive literature review and from information obtained through personal communications with individuals involved with the development of the original MIL-B-5087 specification. The 2.5-mΩ bonding requirement arose not from a theoretically or empirically derived basis but from a practical standpoint and as a means for enforcing proper bond preparation and assembly. The relationship between dc resistance and shielding effectiveness has been investigated with the use of a specialized test joint filled with conductive sealants of varying resistivities to obtain the desired range of dc resistance values. The effect of the test joint geometry, in particular the effect of the plate separation distance, on shielding effectiveness was also evaluated. Although the data collected is by no means all encompassing, it strongly suggest that relaxing the 2.5-mΩ requirement could seriously degrade the EMI/EMP hardness of military aircraft and weapon systems.

7.4 Proposed Changes to Standards, Specification and Handbooks

The authors have developed proposed changes to military standards, specifications and handbooks to reflect the finding of the experimental investigations. The implementation of changes to military standards, specifications and handbooks is a lengthy and complex process requiring inputs and approvals from numerous groups and agencies. For this reason, the results of this task are proposed changes to the appropriate documents, which can then be used as inputs to any modification proceeding. Each recommended change is in a format suitable for direction inclusion in the applicable document.

A larger number of military standards, specifications and handbooks were reviewed to determine appropriate documents for consideration on this task. Table 7.1 is a list of documents which were considered pertinent for modification on this task. The eight documents identified with asterisks were selected as being most relevant for developing proposed changes. These eight documents were later finalized through mutual agreement with Warner Robins ALC/MMEMC. Once confirmation was received from WR-ALC/MMEMC, proposed changes to the mutually agreed upon documents were formulated. The findings and results of the other program tasks were used in the formulation of proposed modifications.

Table 7.1. Documents under consideration for modification.

Document Identifier	Document Title
MIL-B-5087B*	Bonding, Electrical, and Lightning Protection for Aerospace Systems
MIL-STD-188-124A*	Grounding, Bonding and Shielding
MIL-STD-1310D	Shipboard Bonding, Grounding, and Other Techniques for Electromagnetic Capability and Safety Shielding
MIL-STD-1542*	Electromagnetic Compatibility (EMC) and Grounding Requirements for Space System Facilities
MIL-STD-1857	Grounding, Bonding, and Shielding Design Practices
MIL-STD-454J*	Standard General Requirements for electronic Equipment
MIL-STD-462	Electromagnetic Interference Characteristics, Measurement of
MIL-HDBK-253	Guidance for the Design and Test of Systems Protected Against the Effects of Electromagnetic Energy
MIL-E-6051D	Electromagnetic Compatibility Requirements, Systems
NATO STANAG	Bonding and In-Flight Lightning Protection for Aircraft
MIL-STD-1541*	Electromagnetic Compatibility Requirements for Space Systems
AFSC DH 1-4*	Electromagnetic Compatibility
MIL-HDBK-335	Management and Design Guidance, Electromagnetic Radiation Hardness for Air Launched Ordinance Systems
DARCOM-P 706-410	Engineering and Design Guidance Handbook Compatibility
NAVAIR AD 1115	Electromagnetic Compatibility Design Guide for Avionics and Related Ground Support Equipment
DCA NOTICE 310-70-1	DCS Interim Guidance on Grounding Bonding, and Shielding
MIL-HDBK-419	Grounding, Bonding, and Shielding for Electronic Equipments and Facilities

*Documents for Which Modifications Were Proposed

Based on the results obtained from the experimental investigations, a relaxation of the 2.5-mΩ requirement is the empirical study of the relationship between dc resistance and shielding effectiveness indicate that relaxation of the bonding resistance required to 10 mΩ could result in a 20–30 dB reduction in shielding effectiveness (relative to a 2.5 mΩ bond) and that relaxation of the bonding resistance to 100 mΩ could result in 40-55 dB reduction in shielding effectiveness. Furthermore, based on the empirical study of the effects of weathering on electrical performance of the bond, conductive sealants tested from two different vendors (Chomerics silver-coated aluminum-filled polysilicone RTV and Products Research stainless steel-filled polysulfide) maintained a bonding resistance well under the 2.5 mΩ requirement with little or no corrosive performance degradation after 1000 hours in the salt fog environment per ASTM B117. Consequently, relaxation of the 2.5 mΩ requirement was not recommended as part of the proposed modifications.

The proposed modification is presented here one document at a time. Within each document, the current version of the paragraph or section requiring modification is first presented. Immediately following current version is the proposed rewording or modification to the paragraph or section.

7.4.1 Modifications to MIL-B-5087B

MIL-B-5087B, "Military Specification—Bonding, Electrical and Lightning Protection for Aerospace Systems," specifies the characteristics, application and testing of lightning protection and electrical bonding for aerospace systems as well as bonding for the installation and interconnection of electrical and electronic equipment with aerospace systems. The following modifications to this document are recommended:

Current version:	"2.1 The following documents, of the issue in effect on date of invitation for bids or request for proposal, form a part of this specification to the extent specified herein:..."
Proposed modification:	Add MIL-STD-810 to the list of military standards immediately following MIL-STD-143.
Current Version	"3.1.4 Bonding surface preparation—Surface preparation fro and electrical bond...Chemical cleaning and surface preparation shall be in accordance with standard practice (see 6.4)."
Proposed Modification	"3.1.4 Bonding surface preparation.—Surface preparation for and electrical bond...Chemical cleaning and surface preparation shall be in accordance with standard practice (see 3.4.6 and 6.4)."
Current Version	"3.4.1 Bonding installations. - Bonding installations are considered as being permanent and inherently bonded when utilizing metal-t-metal joints by welding, brazing, seating, or swaging. Insulating finishes need not be re-

	moved to comply with 3.1.4 if the resistance requirement is met without such removal. Examples of..."
Proposed Modification	"3.4.1 Bonding installations. - Bonding installations are considered as being permanent and inherently bonded when utilizing metal-to-metal joints by welding, brazing, sweating, or swaging. Examples of..."
Current Version	"3.4.2 Bonding connections. - Bonding connections shall be so installed that vibration, expansion, contraction, or relative movement... The following conditions shall also apply: (a) Parts shall be bonded... (b) Shielding wire ground shall be... (c) Bonding jumpers shall be installed... (d) Bonding connections shall not be... (e) Bonds on plumbing lines shall not be... (f) Current returns and bonds..."
Proposed Modification	"3.4.2 Bonding connections. - Bonding connections shall be so installed that vibration, expansion, contraction, or relative movement... The following conditions shall also apply: (a) Parts shall be bonded... (b) Shielding wire ground shall be... (c) Bonding jumpers shall be installed... (d) Bonding connections shall not be... (e) Bonds on plumbing lines shall not be... (f) Current returns and bonds... (g) Bonding connections subject to corrosive environments shall be protected using appropriate corrosion prevention methods (see 3.4.6)."
Current Version	"3.4.3 Parts impractical to bond with jumpers. - The use of conductive epoxy resins is permitted, provided they conform to the performance requirements of this specification..."
Proposed Modification	"3.4.3 Parts impractical to bond with jumpers. - The use of conductive epoxy resins or conductive sealants is permitted, provided they conform to the performance requirements of this specification..."
Current Version	Currently, there is no specific paragraph in Section 3.4 (Methods) which deals directly with corrosion prevention methods.
Proposed Modifications	In Section 3.4, add the following paragraph 3.4.6 and increment the numbering of succeeding paragraphs appropriately (i.e., current paragraph 3.4.6 becomes paragraph 3.4.7 and so forth).

"3.4.6 Corrosion Prevention Methods. - If aluminum alloys are to be bonded and later be exposed to a corrosive environment, a conductive sealant shall be applied to the faying surfaces before assembly. The conductive sealant shall have sufficient conductivity to meet the applicable electrical requirements of this specification. In addition, the conductive pigments (fillers) used in the sealants shall be compatible with the aluminum alloy substrate material. Silver-coated aluminum or stainless steel fillers are acceptable, whereas silver fillers are not permitted. Other conductive sealant types shall be tested in accordance with MIL-STD-810 and approved by the procuring agency."

Current Version   Currently, there is no specific paragraph in Section 6.4 (Preparation of electrical mating surfaces) which deals directly with preparation of bonding surfaces when the materials are to be subjected to a corrosive environment.

Proposed Modification   Change number of current paragraph 6.4.3.1 to 6.4.3.1.1, paragraph 6.4.3.2 to 6.4.3.1.2, and 6.4.33 to 6.4.3.1.3. Then add the following two paragraphs:
"6.4.3.1 Materials not subject to a corrosive environment."
"6.4.3.2 Materials subject to a corrosive environment - The procedures for preparation of the bond given in 6.4.3.1 are applicable. However, before assembly, a conductive sealant shall be applied to the mating surfaces. The procedures for assembly of the bond given in 6.4.3.1 shall then be followed. Selection of the conductive sealant shall be in accordance with 3.4.6."

Current Version   Currently, none of the figures depicting typical bond connections mention the use of conductive sealants.

Proposed Modification   Figures 1–5, 9 and 10:   Add the note "APPLY CONDUCTIVE SEALANT IF APPLICABLE: immediately below the note "CLEAN TO BASE METAL."
Figure 11: Change last note from "CORROSION PROTECTION SHALL BE AS SPECIFIED IN 3.4.5" to "CORROSION PROTECTION SHALL BE AS SPECIFIED IN 3.4.5 AND 3.4.6."

7.4.2 Modifications to MIL-STD-188-124A

MIL-STD-188-124A, "Military Standard -- Grounding, Bonding and Shielding," establishes the minimum basic requirements and goals for grounding, bonding and shielding of ground-based telecommunications communications-electronics equipment installations, subsystems and facilities including buildings and structure supporting tactical and long haul military communication systems. The following modifications to this document are recommended:

Current Version

"2.1.1 Specifications, Standards, and Handbooks. Unless otherwise specified, the following specifications, standards and handbooks of the issue listed in that issue of the Department of Defense Index of Specifications and Standards (DODISS) specified in a solicitation form in a part of this standard to the extent specified herein..."

Proposed Modification

Add MIL-STD-810 to the list of military standards immediately following MIL-STD-463.

Current Version

"5.1.1.1.6 Other Underground Metals.   Underground metallic pipes entering the facility shall... Adequate corrosion prevention measures shall be taken. Structural pilings, tanks and other..."

Proposed Modification

"5.1.1.1.6 Other Underground Metals.   Underground metallic pipes entering the facility shall... Adequate corrosion prevention measures shall be taken (see 5.2.3.1). Structural pilings, tanks and other..."

Current Version

"5.1.1.3.4 Bonding. All bonds between elements of the lightning protection subsystems shall be made by welding or brazing of UL approved high compression clamping devices. Welding or brazing..."

Proposed Modification

"5.1.1.3.4 Bonding. All bonds between elements of the lightning protection subsystems shall be made by welding or brazing of UL approved high compression clamping devices. Adequate bond protection measures shall be taken (see 5.2.3). Welding or brazing..."

Current Version

"5.2.3.1 Corrosion Protection. Each bonded joint shall be protected against corrosion by assuring that the metals be bonded are galvanically compatible. Bonds shall be painted with a moisture proof paint conforming to the requirements of Federal Specification TT-P-1757 or shall be sealed with a silicon or petroleum-based sealant to prevent moisture from reaching the bond area. Bonds which are located in areas not reasonably accessible for maintenance shall be sealed with permanent waterproof compounds. Iridited or other similarly protected bonds shall not require painting to meet the requirements of this standard."

Proposed Modification

"5.2.3.1 Corrosion Protection. Each bonded joint shall be protected against corrosion by assuring that the metals be bonded are galvanically compatible. Bonds shall be painted with a moisture proof paint conforming to the requirements of Federal Specification TT-P-1757 or shall be sealed with a silicon or petroleum-based sealant to prevent moisture from reaching the bond area. Bonds which are located in areas not reasonably accessible for maintenance shall be sealed with permanent waterproof compounds."

Current Version

"5.2.4 Bond Resistance. All bonds for ground conductors whose primary function is to provide a path for power, control or signal currents and lightning protection shall have a maximum dc resistant of 1 milliohm (0.001 ohm). The resistance across joints or seams in metallic members required to provide electromagnetic shielding shall also be 1 milliohm or less."

Proposed Modification

"5.2.4 Bond Resistance. All bonds for ground conductors whose primary function is to provide a path for power, control or signal currents and lightning protection shall have a maximum dc resistant of 1 milliohm (0.001 ohm). The resistance across joints or seams in metallic members required to provide electromagnetic shielding shall also be 2.5 milliohm or less."

Current Version

5.2.6.5 Bolting. All bonds utilizing bolts and other threaded fasteners shall be... Particular care shall be taken to provide adequate corrosion prevention to all electrical bonds made with bolts and other thread fasteners."

Proposed Modification

5.2.6.5 Bolting. All bonds utilizing bolts and other threaded fasteners shall be... Particular care shall be taken to provide adequate corrosion prevention to all electrical bonds made with bolts and other thread fasteners."

Current Version

In Paragraph 5.2.8.5 (Completion of Bond), no distinction is made between bonds subjected to a corrosive environment and bonds not subjected to a corrosive environment.

Proposed Modification

Add Paragraphs 5.2.8.5.1 and 5.2.8.5.2 as follows:

"5.2.8.5.1 Bonds Subjected to a Corrosive Environment. If aluminum alloys are to be bonded and later be exposed to a corrosive environment, a conductive sealant shall be applied to the faying surfaces before assembly. The conductive sealant shall have sufficient conductivity to meet the applicable electrical requirements of this standard. In addition, the conductive pigments (fillers) used in the sealants shall be compatible with the aluminum alloy substrate material. Silver-coated aluminum or stainless steel fillers are acceptable, whereas silver fillers are not permitted. Other conductive sealant types shall be tested in accordance with MIL-STD-810 and approved by the procuring agency."

"5.2.8.5.2 Bonds Not Subjected to a Corrosive Environment. Mating surface may be joined without the use of the conductive sealants provided that approval is granted by the contracting officer's technical representative."

Current Version

"5.2.11 Connector Mounting. Standard MS-type connectors as well as other shell-type connectors and coaxial connectors shall be mounted so that intimate metallic contact is maintained with the panel on which mounted. Bonding shall be accomplished..."

Proposed Modification

"5.2.11 Connector Mounting. Standard MS-type connectors as well as other shell-type connectors and coaxial connectors shall be mounted so that intimate metallic contact is maintained with the panel on which mounted. Bonding shall be accomplished..." (Note that the term "Metallic" is replaced with the term "electrical" so as not to preclude the use of conductive sealants.).

7.4.3 Modifications to MIL-STD-1541

MIL-STD-1541, "Military Standard -- Electromagnetic Compatibility Requirements for Space Systems" established the EMC requirements for space systems, including launch vehicles, space vehicles, ground systems and associated aerospace ground equipment but does not apply to facilities which house such items. The following changes to this document are recommended:

Current Version

"4.6.2 Case Shielding. Mechanical discontinuities in the case (such as covers, inspection plates and) shall be minimized. All necessary discontinuities critical of rf interference/susceptibility requirements shall be electrically continuous across the interface of the discontinuity to provide a low-impedance current path. Bare or conductively finished metal-to-metal contact, multiple-point spring-loaded contacts or compressive metallic gasketing are desirable methods of obtaining this low-impedance continuity. Ventilation openings shall be designed so as not to degrade the required case shielding effectiveness. Electrical bonding shall be provided where access doors or cover plates form a part of the shielding design. Hinges are not satisfactory conductive paths."

Proposed Modification

"4.6.2 Case Shielding. Mechanical discontinuities in the case (such as covers, inspection plates and) shall be minimized. All necessary discontinuities critical of rf interference/susceptibility requirements shall be electrically continuous across the interface of the discontinuity to provide a low-impedance current path. Bare or conductively finished metal-to-metal contact, multiple-point spring-loaded contacts or compressive metallic gasketing are desirable methods of obtaining this low-impedance continuity. For shields exposed to a corrosive environment, conductive sealants which are galvanically compatible with the substrate materials(s) shall be used to maintain a low-impedance current path over an extended period of time (see MIL-B-5087B). Ventilation openings shall be..."

Current Version

"4.6.3.1 Equipment. All electrical/electronic equipment, including metallic conduit or other conducting items, shall be installed so that there will be a continuous, low-impedance path from the equipment enclosure to the basic structure. This low-impedance path shall have a maximum direct current (dc) bonding resistance of 2.5 milliohms across the mated surfaces as demonstrated by contractor test. Bonding shall be accomplished by bare, clean, metal-to-mental contact of all mounting plate, rack, shelf, bracket and structure mating surface to form a continuous, low-impedance bond from the equipment mounting plates. Bare metal surfaces when subject to corrosion shall be protected by conductive surface finishes such as Alodine or Iridite. Suitable jumpers shall be used across any necessary vibration isolation mounts or heat sink plates."

Proposed Modification

"4.6.3.1 Equipment. All electrical/electronic equipment, including metallic conduit or other conducting items, shall be installed so that there will be a continuous, low-impedance path from the equipment enclosure to the basic structure. This low-impedance path shall have a maximum direct current (dc) bonding resistance of 2.5 milliohms across the mated surfaces as demonstrated by contractor test. All mounting plate, rack, shelf, bracket and structure mating surfaces all be prepared by removing any grease, oil or other non-conductive films to provide clean, bare metal surface for bonding. Metal surfaces subject to corrosion shall be protected using the corrosion prevention methods specified in MIL-B-5087B. Suitable jumpers shall be used across any necessary vibration isolation mounts or heat sink plates."

7.4.4 Modifications to MIL-STD-1542

MIL-STD-1542, "Military Standard -- Electromagnetic Compatibility (EMC) and Grounding Requirements for space System Facilities," established the general EMC and grounding requirements for space system ground facilities, including structures that house electrical/electronic devices or equipment, such as service structures, tracking station buildings, satellite control rooms, computer rooms and space craft or booster assembly buildings. The following modification to this document are recommended:

Current Version	Section 5 (Detailed Requirements) does not include a discussion of corrosion prevention methods.
Proposed Modification	Add Paragraph 5.3 at the end of Section 5 as follows: "5.3 Corrosion Protection. All electrical bonds subject to corrosion shall be protected using the corrosion prevention methods specified in MIL-B-5087B."

7.4.5 Modifications to MIL-STD-454K

MIL-STD-454K, "Military Standard -- Standard General Requirements for Electronic Equipment," is a technical standard for the design and fabrication of electronic equipment for the Department of Defense. Section 5, Electrical, of Requirement 1, Safety (Personnel Hazard), provides dc resistance values for bonding inconsistent with those specified in MIL-B-5087B, even through this document is reference in Requirement 1. Consequently, the following two paragraphs should be modified:

Current Version	"5.3.1 Hinged or slide mounted panels and doors. Panels and doors... A ground shall be considered satisfactory if the electrical connection between the door or panel and the system tie point exhibits a resistance of 0.01 ohm or less and has..."
Proposed Modification	"5.3.1 Hinged or slide mounted panels and doors. Panels and doors... A ground shall be considered satisfactory if the electrical connection between the door or panel and the system tie point exhibits a resistance of 0.01 ohm or less and has..."
Current Version	"5.4 Grounding to chassis. Ground connection to an electrically conductive chassis or frame shall be... When aluminum or aluminum alloys are used, the metal around the grounding screw or bolt may be covered with a corrosion resistant film only if the resistance through the film is not more than 0.002 ohm."
Proposed Modification	"5.4 Grounding to chassis. Ground connection to an electrically conductive chassis or frame shall be... When aluminum or aluminum alloys are used, the metal around the grounding screw or bolt may be covered with a corrosion resistant inhibiting conductive sealant if the resistance through the film or sealant is less than or equal to 2.5 milliohms."

7.4.6 Modifications to MIL-HDBK-253

MIL-HDBK-253, "Guidance for the Design and Test of Systems Protected against the Effects of Electromagnetic Energy," provides program managers with EMI design guidance to ensure that systems are designed to operate and survive in anticipated tactical electromagnetic environments. The following modifications to this handbook are recommended:

Current Version

"8.1.2.2 Cable Shielding. There are several methods for shield cables as discussed in (a) through (d):
(a) Braid...
(b) Conduit... Degradation of shielding conduit is usually not because of insufficient shielding properties of the conduit material but rather the result of discontinuities in the cable shield. These discontinuities usually result from splicing or improper termination of the shield.
(c) Solenoids...
(d). The..."

Proposed Modification

"8.1.2.2 Cable Shielding. There are several methods for shield cables as discussed in a through d:
(a) Braid...
(b) Conduit... Degradation of shielding conduit is usually not because of insufficient shielding properties of the conduit material but rather the result of discontinuities in the cable shield. These discontinuities usually result from splicing or improper termination of the shield. Discontinuities may also result from corrosion of improperly protected bonds.
(c) Solenoids...
(d) The..."

Current Version

"8.1.3 Summary of shielding practices. The following represent what might be considered the more salient points on shielding design considerations:
(a) Good...
(b) Magnetic...
(c) Any...
(d) In...
(e) Multiple...
(f) All openings or discontinuities should be treated in the design process to assure minimum reduction in shield effectiveness. Particular attention should be paid to selection of materials that are not only suitable for shielding, but from the electrochemical corrosion viewpoint as well.
(g) When..."

(h) Surfaces...
(i) Conductive...
(j) Shielding...

Proposed Modification   "8.1.3 Summary of shielding practices. The following represent what might be considered the more salient points on shielding design considerations:

(a) Good...
(b) Magnetic...
(c) Any...
(d) In...
(e) Multiple...
(f) All openings or discontinuities should be treated in the design process to assure minimum reduction in shield effectiveness. Particular attention should be paid to selection of materials that are not only suitable for shielding, but from the electrochemical corrosion viewpoint as well. The incorporation of compatible materials and proper corrosion prevention measures, particularly at bonded joints and seams, is critical in terms of maintaining the integrity of the shield for the life of the system.
(g) When...
(h) Surfaces...
(i) Conductive...
(j) Shielding...

Current Version   "8.2.2.1 Surface treatment. Both direct and indirect bonding connections required metal-to-metal contact of bare surfaces with the area cleaned for bonding being slightly larger than the area to be bonded. Ridges of paint... Where aluminum of its alloys are used, corrosion resistant finished that offer low electrical resistance are available."

Proposed Modification   "8.2.2.1 Surface treatment. Both direct and indirect bonding connections required metal-to-metal contact of bare surfaces, except where a suitable conductive sealant is used. The area cleaned for bonding should be slightly larger than the area to be bonded. Ridges of paint... Where aluminum or its alloys are used, corrosion resistant finished that offer low electrical resistance are available."

Current Version   "8.2.3 Summary of bonding design guidelines. The effectiveness of a bond depends on... Many examples of the variety of techniques available for low impedance connections for bonds are available in MIL-STD-1310 and IL-B-5087. Some general guidelines for obtaining good bonds are provided in (a) through (g):

(a) The secret to good bonding is intimate contact between metal surfaces. Surfaces must be smooth and clean and not coated with a nonconductive finish. The fastening method must exert sufficient pressure to hold the surfaces in contact in the presence of deforming stresses, shock and vibrations associated with the equipment and its environment.

(b) Bonds are always best made by joining similar metals. If this is not possible, special attention must be paid to the possibility of bond corrosion through the choice of the materials to be bonded, the selection of supplementary components (such as washers) that will affect replaceable elements only, and the use of protective finished.

(c) Solder...

(d) Protection of the bond from moisture and other corrosion effects must be provided where necessary.

(e) Bonding...

(f) Jumpers...

(g) It..."

Proposed Modification   "8.2.3 Summary of bonding design guidelines. The effectiveness of a bond depends on... Many examples of the variety of techniques available for low impedance connections for bonds are available in MIL-STD-1310 and IL-B-5087. Some general guidelines for obtaining good bonds are provided in (a) through (g).

(a) The secret to good bonding is intimate electrical contact between metal surfaces, either through direct physical contact or through a conductive sealant applied between the surfaces. Surfaces must be smooth and clean and not coated with a nonconductive finish. The fastening method must exert sufficient pressure to hold the surfaces in contact in the presence of deforming stresses, shock and vibrations associated with the equipment and its environment.

(b) Bonds are always best made by joining similar metals. If this is not possible, special attention must be paid to the possibility of bond corrosion through the choice of the materials to be bonded, the selection of supplementary components (such as washers) that will affect replaceable elements only, and the use of protective finished.

(c) Solder...

(d) Protection of the bond from moisture and other corrosion effects must be provided where necessary. If aluminum or aluminum alloys are to be bonded and later be exposed to a corrosive environment, a conductive sealant should be applied to the faying surfaces before assembly. The conductive sealant must have sufficient conductivity to meet the applicable electrical requirements, and the conductive pigments (fillers) used in the sealants must be compatible with the aluminum substrate material. Silver-coated aluminum or stainless steel fillers are acceptable whereas silver fillers should be avoided.

(e) Bonding...

(f) Jumpers...

(g) It..."

7.4.7 Modifications to AFSC Design Handbook 1-4

AFSC Design Handbook 1-4, "Electromagnetic Compatibility," provides system designers with electromagnetic compatibility design principles, information, guidance and criteria and is intended to serve as a central source of electromagnetic compatibility design data. The following modifications to this handbook are recommended:

Current Version

"DESIGN NOTE 5D1 SUB-NOTE 4 (1). Guidance for Electrical Bonding and Grounding:

1. Clean...
2. Weld...
3. Where protective films are absolutely required, ensure that the film material is a good d conductor. Some suitable protective films are: silver or gold plating or other plated metals of good conductivity (oakite #36, alodine #1000, iridite #14 and iridite #18P).
4. Ensure...
5. If the surfaces are not inert in their storage and operating environments, provide surface protection according to rule "3" or take other suitable measures to ensure the maintenance of the bond for the service life of the equipment.
6. Do not...
7. Do not...
8. Consider...
9. Compress...
10. Ensure...
11. Check...
12. Ensure...
13. Use...

Proposed Modification   "DESIGN NOTE 5D1 SUB-NOTE 4 (1). Guidance for Electrical Bonding and Grounding.
1. Clean...
2. Weld...
3. Where protective films are required, ensure that the film material is a good conductor. Some suitable protective films are: silver or gold plating or other plated metals of good conductivity (oakite #36, alodine #1000, iridite #14 and iridite #18P).
4. Ensure...
5. If the surfaces are not inert in their storage and operating environments, provide surface protection according to rule "3" or take other suitable corrosion prevention measures (see MIL-B-5087) to ensure the maintenance of the bond for the service life of the equipment.
6. Do not...
7. Do not...
8. Consider...
9. Compress...
10. Ensure...
11. Check...
12. Ensure...
13. Use...

Current Version   "DESIGN NOTE 5D2.1 INTRODUCTION Bonding is defined herein as the establishment of the lowest obtainable resistance between two conducting surfaces. The following conditions are necessary to achieve the lowers resistance between surfaces:
(a) Use no films or layers or material between the surfaces unless the film material is better conductor than the materials being bonded, and
(b) The mating surfaces must be smooth and contoured so that maximum surface area is in actual contact."

Proposed Modification   "DESIGN NOTE 5D2.1 INTRODUCTION Bonding is defined herein as the establishment of the lowest obtainable resistance between two conducting surfaces. The following conditions are necessary to achieve a low impedance between surfaces and to maintain this low impedance over the operational lifetime of the system, subsystem or equipment:
(a) All finished, sealants and materials used between the surfaces must be of comparable conductivity to that of the materials being bonded,
(b) The mating surfaces must be smooth and contoured so that maximum surface area is in actual contact."

(c). Adequate corrosion protection must be provided (see DN 5D3.6)."

Current Version

"DESIGN NOTE 5D3.4 JOINT SURFACE TREATMENT AND SEALING Specify additional finishes, such as paint or plating, with caution. Finishing of the anodic material alone may produce severe corrosion at any finish imperfection. When dissimilar metals are in contact, do not cover the surface of only the anodic material; either cover the surface of both metals or only the cathode (see SN 4(1)). This is again due to the unfavorable anode to cathode area ratio."

Proposed Modification

"DESIGN NOTE 5D3.4 JOINT SURFACE TREATMENT AND SEALING Protection of the bond from moisture and other corrosion effects must be provided where necessary. If aluminum or aluminum alloys are to be bonded and later be exposed to a corrosive environment, a conductive sealant should be applied to the faying surfaces before assembly. The conductive sealant must have sufficient conductivity to meet the applicable electrical requirements, and conductive pigments (fillers) used in the sealants must be compatible with the aluminum substrate material. Silver-coated aluminum or stainless steel fillers are acceptable whereas silver filers should be avoided. Specify additional finishes, such as paint of plating, with caution. Finishing of the anodic material alone may produce severe corrosion at any finish imperfection. When dissimilar metals are in contact, do not cover the surface of only the anodic material; either cover the surface of both metals or only the cathode (see SN 4(1)). This is again due to the unfavorable anode to cathode area ratio."

Current Version

"DESIGN NOTE 5F1.2 INTERFACES If it were not for the many mechanical and electrical interfaces required in an aerospace system, the shielding problem would be reduced to choosing a proper shield material and apply a simple box concept. Since each interface degrades the shield to some degree, the selection and implementation of techniques to provide continuity at these interfaces is important. Sub-Note 2(1) illustrates some of these interfaces."

Proposed Modification

"DESIGN NOTE 5F1.2 INTERFACES If it were not for the many mechanical and electrical interfaces required in an aerospace system (some of which are illustrated in Sub-Note 2(1)), the shielding problem would be reduced to choosing a proper shield material and applying a simple box concept. Since each interface degrades the shield to some degree, the selection and implementation of techniques to provide continuity at these interfaces is important. In addition, the incorporation of proper interfaces is important. In addition, the incorporation of proper corrosion prevention measures is important in maintain the integrity of the shield over the life of the system."

Current Version

"Design NOTE 5F1 SUB-NOTE 4(1) Guidance for Shielding
1.   Design...
2.   Use...
3.   Use...
4.   Compress...
5.   Use...
6.   Check...
7.   Electrically...
8.   Whenever...
9.   Ensure...

Proposed Modification

"Design NOTE 5F1 SUB-NOTE 4(1) Guidance for Shielding
1.   Design...
1.   Design...
2.   Use...
3.   Use...
4.   Compress...
5.   Use...
6.   Check...
7.   Electrically...
8.   Whenever...
9.   Ensure...
10.   Ensure that adequate corrosion prevention measures are applied to maintain shielding effectiveness of joints and seams."

Current Version

"DESIGN NOTE 5F5.2.1 OVERLAPPING SEAM An acceptable alternative technique is the overlap seam shown in "D" of SN 2(1). In an overlap seam, all nonconductor materials must be removed from the mating surfaces before the surfaces are crimped, and the crimping must be performed under sufficient pressure to ensure positive contact between all mating surfaces. Sub-Note 2.1(1) summarizes, in order of preference, techniques for implementing permanent of semipermanent seams."

Proposed Modification

"DESIGN NOTE 5F5.2.1 OVERLAPPING SEAM An acceptable alternative technique is the overlap seam shown in "D" of SN 2(1). In an overlap seam, all nonconductive materials must be removed from the mating surfaces before the surfaces are crimped, and the crimping must be performed under sufficient pressure to ensure positive contact between all mating surfaces. Conductive sealants may be applied before crimping to provide corrosion protection while maintaining an effective bond. Sub-Note 2.1(1) summarizes, in order to preference, techniques for implementing permanent of semipermanent seams."

Current Version

"DESIGN NOTE 5F5.2.3 CONSIDERATION Seams... To assure adequate and properly implemented bonding techniques, observed the following recommendations:
(a) All...
(b) All...
(c) When protective coatings are necessary, design them so that they can be easily removed from mating surfaces. Since the mating of bare metal to bare metal is essential for a satisfactory bond, a conflict may arise between the bonding and finish specification. It is preferable to remove the finish where compromising of the bonding effectiveness would occur.
(d) Generally...
(e) Mating...
(f) The...
(g) Bolted...
(h) When...
(i) When pressure bonds are made, the surfaces must be clean and dry before mating and then held together under high pressure to minimize the growth of oxidation due to moisture entering the joint, since the joint may not be 100% moisture-tight. The periphery of the exposed joint should then be sealed with a suitable protective compound and, whenever possible, one that is highly conductive to RF currents."

Proposed Modification

"DESIGN NOTE 5F5.2.3 CONSIDERATION Seams... To assure adequate and properly implemented bonding techniques, observed the following recommendations:
(a) All...
(b) All...
(c) When protective coatings are necessary, design them so that they can be easily removed from mating surfaces. Since the mating of bare metal to bare metal is desirable for a satisfactory bond, a conflict may arise between the bonding and finish specification. Conductive sealants may be used in the mating surface provided that they are verified though testing to meet the bonding and corrosion requirements. Any

compromise between the bonding and corrosion/finish requirements, if necessary, should be approved by the procuring agency.

(d) Generally...

(e) Mating...

(f) The...

(g) Bolted...

(h) When...

(i) When pressure bonds are made, the surfaces must be clean and dry before mating and then held together under high pressure to minimize the growth of oxidation due to moisture entering the joint. Since the joint may not be 100% moisture-tight, a conductive sealant may be useful in excluding moisture while maintaining a high conductivity between the mating surfaces. The periphery of exposed, untreated joints should be sealed with a suitable protective compound and, whenever possible, one that is highly conductive to RF currents."

Current Version

"DESIGN NOTE 5F5.2.4 RF BONDS When implementing bonding techniques, always remember that bonding straps do not provide a low impedance current path at RF frequencies. The important impedance exists at radio frequencies. There is little correlation between the dc resistance of a bond and its RF impedance. Even the measured RF impedance of bonds, such as jumpers, straps or rivets, is not a reliable indication of the bonding effectiveness in he actual installation. Conductive epoxies and pastes do not always produce sufficient RD bonds. Even when proved effective in given instances, they have been know to degrade RF shielding effectiveness under conditions of strain, pressure and the passage of time."

Proposed Modification

"DESIGN NOTE 5F5.2.4 RF BONDS When implementing bonding techniques, always remember that bonding straps do not necessarily provide a low impedance current path at RF frequencies. The RF impedance of a bond is more important than its dc resistance in determining overall bonding/shielding effectiveness. There is little correlation between the dc resistance of a bond and its RF impedance. Further, the measured RF impedance of bonds, such as jumpers, straps or rivets, is not a reliable indication of the bonding effectiveness in the actual installation. Conductive epoxies and pastes do not always produce sufficient RF bonds and, therefore, should be demonstrated by test to maintain RF shielding effectiveness under conditions of strain, pressure, corrosive environments and the passage of time."

Current Version   DESIGN NOTE 5H1 SUB-NOTE 1(1) EMI Summary List" (forth column, fifth paragraph from the bottom) "Have dissimilar metals been used in the enclosure? In this compatible with the expected environment?"

Proposed Modification   DESIGN NOTE 5H1 SUB-NOTE 1(1) EMI Summary List" (forth column, fifth paragraph from the bottom) "Have dissimilar metals been used in the enclosure? In this compatible with the expected environment? Have adequate corrosion prevention measures been taken for maintaining the shield integrity?'

7.4.8 Modifications to NATO STANAG No. 3659

NATO Standardization Agreement (STANAG) No. 3659, "Bonding and In-Flight Lighting Protection for aircraft,'" established the aircraft bonding and in-flight lightning protection requirements and associated tests for participating nations. The following modifications to this Standardization Agreement are recommended:

Current Version   "11. BONDING METHODS
(b) Bonding installations. Bonding...
(c) Bonding connections. Bonding... The following conditions shall also apply:
    (1) Parts...
    (2) Shielded...
    (3) Bonding...
    (4) Bonding...
    (5) Bonds...
    (6) For....
(d) Parts impractical to bond with jumpers. The used of conductive epoxy resins in permitted, provided they conform to the performance requirements of this specification...
(e) Circular conductors. Bonding...
(f) Dissimilar metals. When...
(g) Refinishing. When...
(h) Intermittent electrical contact. Bonding...
(i) Unapproved bonding methods. Anti-friction..."

Proposed Modification   "11. BONDING METHODS
(j) Bonding installations. Bonding...
(k) Bonding connections. Bonding... The following conditions shall also apply:
    (1) Parts...
    (2) Shielded...
    (3) Bonding...

(4) Bonding...

(5) Bonds...

(6) For....

(7) Bonding connections subject to corrosive environments shall be protected with the appropriate corrosion prevention methods (see paragraph 11f).

(l) Parts impractical to bond with jumpers. The used of conductive epoxy resins or conductive sealants in permitted, provided they conform to the performance requirements of this specification...

(m) Circular conductors. Bonding...

(n) Dissimilar metals. When...

(o) Corrosion Prevention Methods. If aluminum or aluminum alloys are to be bonded and later be exposed to a corrosive environment, a conductive sealant shall be applied to the faying surfaces before assembly. The conductive sealant shall have sufficient conductivity to meet the applicable electrical requirements of this document. In addition, the conductive pigments (fillers) used in the sealants shall be compatible with the aluminum substrate material. Silver-coated aluminum or stainless steel fillers are acceptable whereas silver-filed sealants are not permitted. Other conductive sealant types must be tested in a salt-fog (accelerated corrosion) environment and approved by the procuring agency.

(p) Refinishing. When...

(q) Intermittent electrical contact. Bonding...

(r) Unapproved bonding methods. Anti-friction..."

Current Version

Currently there is no specific paragraph in Annex C (Bonding Surface Preparation) which deals directly with preparation of bonding surfaces when the materials are to be subjected to a corrosive environment.

Proposed Modification

Change number of current paragraph 2.c(1) to 2.c(1) (a) and paragraph 2.c(2) to 2.c(1) (b). Then add the following two paragraphs:

"2.c. (1) Materials not subjected to a corrosive environment."

"2.c.(2) Materials subjected to a corrosive environment. The procedures for preparation of the bond given in 2.c.(1) are applicable. However, before assembly, a conductive sealant shall be applied to the mating surfaces. The procedures for assembly of the bond given in 2.c.(1) shall then be followed. Selection of the conductive sealant shall be in accordance with 11.f."

7.5 Recommendations for Corrosion Prevention Materials/Processes for Use on Existing (and Future) Aircraft and Weapon Systems

High values of shielding effectiveness necessitate that a low impedance exist across a joint or bond. The use of specific conductive sealants provide the capability of maintaining a 2.5 mΩ resistance and comparably low transfer impedance even after 1000 hours of exposure in a Salt Spray (Fog) Chamber per ASTM B117. Therefore, we recommend that the above successful conductive sealants be used on existing and future aircraft where sealants can be liquid applied. We recommend a 2-part polyurethane or polysulfide system excluding silicone materials. Silicone materials are prone to produce surface energy problems during coating operations (e.g., Fisheyes). Where feasible, joints could be disassembled and reassembled with conductive sealant. During manufacture of new aircraft and weapon systems, the conductive sealant could be applied before assembly of joints.

7.6 Recommendations for Using Corrosion Prevention Materials for retrofitting Existing Nuclear Harden Aircraft and Weapons Systems

7.6.1 Retrofitting Aircraft with Conductive Sealant

We recommend that existing aircraft and weapon systems be retrofitted with conductive sealant where feasible. This will include disassembly of joints of these systems and application of liquid-applied conductive sealant. We recommend the use of a 2-part liquid-applied conductive sealant discussed above having survived 1000 hours of Salt Spray Chamber, ASTM B117. Further, we recommend a 2-part polyurethane or polysulfide sealant and recommend omitting polysilicone materials. Also, it is imperative that aircraft and weapon systems that have been in service for a sufficient time to cause corrosion at the joint surfaces be surface cleaned before applying the conductive sealant.

7.6.2. Annodized and Chemical Conversion Type Treated Aluminum Surfaces

The anodized and chemically treated aluminum possesses resistances of about 5 mΩ/in^2 (MIL-C-5541). The resistance across such joints will be the sum of sealant and surface treatment resistances.

7.7 Resealable Joints Using Liquid Applied Conductive Sealants

7.7.1 Resealable Joints in Salt Spray Chamber

The aluminum joints were sealed with Proseal RW-2128-71, after the top place of the joint was coated with a film of poly(dimethylsiloxane) with a viscosity of 60,000 centipoise. Presumably, the release agent will allow the joint to "break-apart" without damaging the joint structure since the sealant alone serves as and adhesive. The resistance across the joint using a release agent was not known, and its measurement was of interest. The joints were exposed in the Salt Fog Chamber for 1000 hours. The resistance and shielding effectiveness were measured in regular intervals, and the results are listed in Table 7.2. The shielding effectiveness data is contained in Appendix H. Shielding effectiveness versus salt fog chamber exposure time is plotted in Fig. 7.1.

Referring to Table 7.2, the average resistance, excluding the control joint, is 0.0289 Ω or 28.95 mΩ, which is 17.02 times (28.95/1.70) greater than the same sealant without release agent on the top plate. The shielding effectiveness was an average of 70.84 db, excluding the control joint.

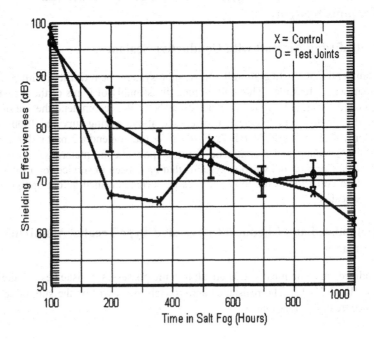

Fig. 7.1. Shielding effectiveness vs. exposure time in salt spray chamber for a resealable joint.

Table 7.2. Resistance and shielding effectiveness of joints sealed with conductive sealant[a] pretreated with poly(dimethylsiloxane) release agent, and exposed to 1000 hours of salt spray chamber.

Joint No.	Resistance (Ω)	Shielding effectiveness (dB)
1	35+	61.68
2	0.0750	67.86
3	0.0300	70.83
4	0.0043	72.78
5	0.0061	71.89
	Ave.[b] = 0.0289	Ave. = 70.84

[a]Proseal RW-2128-71, Products Research Corporation.
[b]Average does not include control joint.

7.7.2 Opening Resealable Joints

Fasteners were removed from the joints, and the top plate was removed by prying with a wedge such as a screw drive. It was possible to remove the plate without severely bending any part of the joint. Also, presoaking the bonded area with toluene enhanced the ease of plate separation.

7.7.3 Comments

Based on the encouraging first set of results from these experiments, we recommend experimenting with other release agents, which are capable of providing greater conductivity and better joint release properties.

7.8 Shielding Effectiveness of Non-Conductive Corrosion Inhibitive Sealant

Aluminum joints were sealed with non-conductive corrosion inhibitive sealant and the shielding effectiveness was measured. The results are listed in Table 7.3, and the shielding effectiveness data are contained in Appendix I. Predictably, the resistance of each joint is so large that it is out of range of the meter (22 Ω). Since the resistance was large, it is expected that the shielding effectiveness will be low, and that correlation is confirmed by the shielding effectiveness values in Table 7.3.

Table 7.3. Resistance and Shielding Effectiveness of Joints Sealed with Corrosion Inhibitive Sealant (Non-Conductive)

Joint No.	Resistance[a] (Ω)	Shielding Effectiveness (dB)
1	22+	42.73
2	22+	44.47
3 (control)	< 0.001	99.31
4	22+	40.41
5	22+	42.19
	Ave.[b] = N/A	Ave. = 42.45

[a]All resistance and shielding effectiveness values measured after 24 hours of curing, no corrosion testing.
[b]Average values do not include control joint.

7.9 Aerospace Material Specification

Based on the above research, development and applications, the Aerospace Material Specification 3262 was sponsored by the U.S. Air Force adopted by the Society of Automotive Engineers, February 2000:

> Society of Automotive Engineers, International
> 400 Commonwealth, Drive, Warrendale, PA 150096-0001
> AMS 3262, SEALING COMPOUND, SILICONE RUBBER, TWO-PART, ELECTRICALLY CONDUCTIVE AND CORROSION RESISTANT, FOR USE FROM –67 TO 500 °F (-55 TO 260 °C)

8

EMI Gaskets

8.1 EMI Gasket Applications and Selection Criteria

EMI gaskets have been in use for many decades, but recent advances hold promise for new applications, increased performance, and enhanced durability. EMI gaskets are used to provide a continuous, low-impedance electrical bond between two metal faying surfaces. They are pliable so as to mold to typical faying surface irregularities. They can be used for a variety of applications including grounding, lightning protection, and electromagnetic shielding.

Selection of an EMI gasket is governed by various performance criteria including: (1) shielding effectiveness or transfer impedance over a specified frequency range, (2) mounting methods and closure forces, (3) galvanic compatibility with the metal substrate material and corrosion resistance from the external environment, (4) operating temperature range, and (5) cost.

EMI gaskets can be grouped according to their material composition and geometrical configuration [10, 11], resulting in five basic groupings. The characteristics of these basic EMI gasket types are discussed in the following paragraphs.

8.2 Stamped and Formed Metal Gaskets

Examples of stamped and formed metal gaskets include stainless steel, phosphor bronze, and BeCu fingerstock as well as spiral wound gaskets. The advantages of these gaskets include robust construction and high shielding effectiveness. BeCu is by far the most popular for a number of reasons. Of all the materials that can be converted into a spring, BeCu has the highest conductivity – more than twice that of such materials as phosphor bronze and stainless steel. Stainless steel offers a low cost alternative to BeCu, but lacks the resilient spring properties of BeCu and is thus not recommended for high-cycling applications. Thus, stainless steel is best suited for use in static joints and limited-access panels. Another advantage of these gaskets is that the sliding contact tends to scrape away surface oxides and other non-conductive contaminates from the faying surface, thus providing improved electrical contact.

The shielding effectiveness of stamped and formed metal gaskets ranges from a high of about 120 dB for BeCu down to about 60-70 dB for stainless steel. Shielding effectiveness tends to fall off at high frequencies, where the slots or openings become an appreciable portion of a wavelength. Compression forces range from 10 – 20 lb per linear foot for standard fingerstock metal thicknesses and profiles. Usable compression heights for BeCu gaskets range from 20 – 80% of their free height dimension. Compression set values range from less than 1% for BeCu up to 25-40% for some grades of stainless steel, provided that the gaskets are operated within the specified range of compression (over compressing any stamped metal gasket greatly increases its compression set).

8.3 Wire Mesh Gaskets

Examples of wire mesh gaskets include tin-plated, copper-clad steel (SnCuFe), Monel (a nickel copper alloy), aluminum, and beryllium copper (BeCu). Monel wire has probably been the most widely used wire mesh gasket type due to its high strength and durability, and good aging properties. SnCuFe offers high shielding effectiveness performance, but poor corrosion resistance limits the utility for many applications. Another wire mesh type is a knitted wire over a sponge or nonconductive elastomer core, which provides the mesh gasket with better spring qualities and provides a superior dust and moisture seal. A recent development has been the BeCu wire mesh, which offers excellent shielding performance and spring characteristics. BeCu can also be plated to improve its galvanic compatibility with specific flange surfaces under harsh environmental conditions.

The shielding effectiveness of wire mesh gaskets ranges from 40–105 dB, depending on the material, construction, and shielding effectiveness measurement technique. Compression load forces range from 5 lb/in^2 for hollow-core gaskets up to 20 lb/in^2 for solid-core gaskets. Hollow-core gaskets achieve a useable contact resistance with about 20% deflection and can be compressed up to 75%, depending on the material. Solid-core mesh gaskets require less compression deflection to function efficiently, but have a maximum deflection of about 40%. Compression set for BeCu wire mesh is less than 1% but reaches 20% for other hollow- or elastomer-core meshes. Solid-core wire mesh gasket compression set can be as high as 30%.

Wire mesh gaskets are most commonly used in static seal or limited access applications due to the limitations in the spring properties of the original materials and the mounting methods available. Wire mesh is typically used in a groove-mount configuration.

8.4 Conductively-Loaded Elastomeric Gaskets

Conductively-loaded elastomeric gaskets consist of a polymeric binder, such as silicone, fluorosilicone, or Ethylene Propylene Diene Monomer (EPDM), that is loaded with conductive fillers, such as silver, copper, carbon, nickel, aluminum, nickel-plated graphite, silver-plated aluminum, silver-plated copper, silver-plated nickel, or silver-plated glass spheres. Conductively-loaded elastomers not only provide EMI shielding, but can also function as a pressure and moisture seal. Military grade elastomers qualified to MIL-G-83528 are manufactured to tight quality guidelines, but tend to be rather expensive. Most manufacturers supplement their military grade products with less expensive, commercial grade products that feature conductive fillers such as silver-plated copper or silver-plated glass spheres. A variation of the conductively-loaded elastomer is the co-extruded version, which features a conductively loaded outer liner over a non-conductive hollow inner core. Another variation is the compound gasket, which typically has two, symmetrical o-ring gaskets. One o-ring, placed on the outside of an enclosure seal, is non-conductive and acts as a dedicated environmental seal to protect against moisture and contaminants. The other o-ring, placed on the inside of an enclosure seal, is conductively-loaded and provides high shielding effectiveness.

The shielding effectiveness of conductively-loaded elastomers ranges from about 60 dB for commercial grade products up to about 120 dB for military-grade products. The shielding effectiveness is dependent on the type and amount of conductive filler used as well as the compression force. Conductively-loaded elastomers require a minimum deflection of 10 –30% to function properly, and their maximum deflection is typically limited to about 50%, dependent on the specific gasket geometry. Compression loads tend to be very high, ranging from about 25 lb per linear foot up to 100 lb per linear inch. Compression set values at 100 °C range from 5% for hollow-core extrusions to 30% or more for some solid-core constructions.

8.5 Conductive Fabric-Clad Foam Gaskets

One of the more recent advances, conductive fabric-clad foam gaskets are made from a metallized fabric, such as silver on nylon, nickel copper on nylon or polyester, or other combination including tin copper or nickel silver. Shielding effectiveness ranges from about 60 to 100 dB, depending on the materials and the quality of the plating process used. These gaskets offer the lowest compression force of any type of gasket, due to their foam cell structures and flexible conductive fabrics. For example, the C- or V-shape profile gaskets provide compression forces as low as 1.3 lbs per linear foot. Fabric-clad EMI gaskets have a usable deflection range of 10 to 75%. Compression set values at 70 °C range from about 5 to 35%, depending on the foam core material used in the construction of the gasket. Open-cell polyurethane foams are generally less susceptible to compression set than their closed-cell counterparts.

8.6   Conductive Gel Gaskets

Conductive gel gaskets, which are typically made from wire mesh impregnated with cured silicone or flourosilicone gel, are useful for EMI shielding, electrical grounding, and environmental sealing. Under pressure, the silicone gel exhibits displacement characteristics, and its ability to adhere to a surface on contact enables it to form an effective moisture and pressure barrier. The material's conformability keeps electrolyte away from the metal faying surfaces, inhibiting corrosion by preventing formation of a galvanic cell. One implementation makes use of a circular compressed Monel (copper-nickel alloy) wire mesh rope made that is impregnated with a cured silicone gel. They are a cost-effective alternative to compound elastomeric gaskets or one-piece, conductively-loaded elastomers. The silicone gels have a −40 to 125 °C temperature range, whereas the flourosilicone gels have a −40 to 150 °C temperature range of operation.

Appendix A:
Shielding Effectiveness Data for Test Joints with Varying Resistances

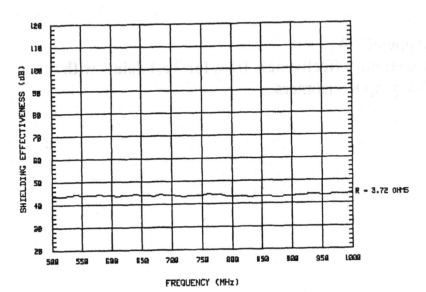

SHIELDING EFFECTIVENESS VERSUS FREQUENCY OF SAMPLE
WITH RESISTANCE GREATER THAN 20 MEGAOHMS,
R - 3.72 OHMS

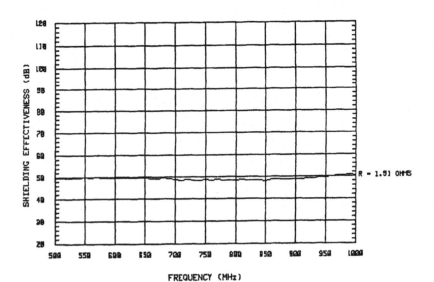

SHIELDING EFFECTIVENESS VERSUS FREQUENCY OF SAMPLE
WITH RESISTANCE GREATER THAN 20 MEGAOHMS,
R - 1.91 OHMS

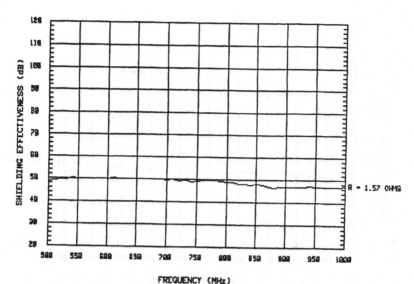

SHIELDING EFFECTIVENESS VERSUS FREQUENCY OF SAMPLE
WITH RESISTANCE GREATER THAN 20 MEGAOHMS,
R - 1.57 OHMS

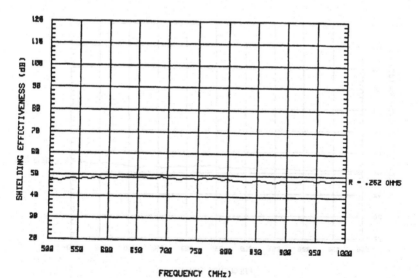

SHIELDING EFFECTIVENESS VERSUS FREQUENCY OF SAMPLE
WITH RESISTANCE GREATER THAN 20 MEGAOHMS,
R - .252 OHMS

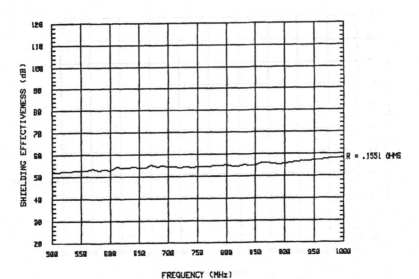

SHIELDING EFFECTIVENESS VERSUS FREQUENCY OF SAMPLE
WITH RESISTANCE GREATER THAN 20 MEGAOHMS,
R = .1551 OHMS

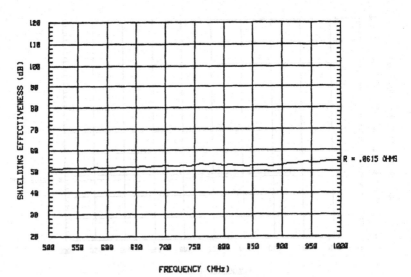

SHIELDING EFFECTIVENESS VERSUS FREQUENCY OF SAMPLE
WITH RESISTANCE GREATER THAN 20 MEGAOHMS,
R = .0615 OHMS

A7

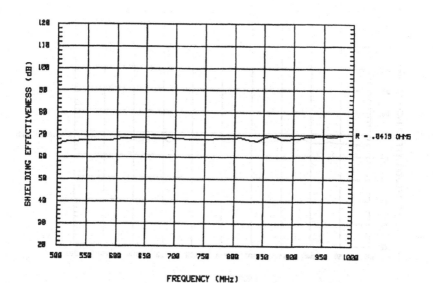

SHIELDING EFFECTIVENESS VERSUS FREQUENCY OF SAMPLE
WITH RESISTANCE GREATER THAN 20 MEGAOHMS,
R - .0419 OHMS

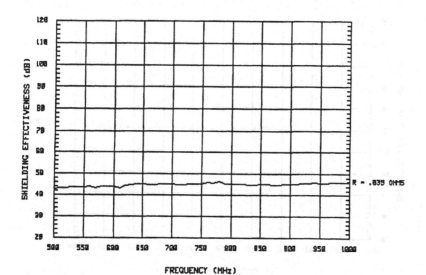

SHIELDING EFFECTIVENESS VERSUS FREQUENCY OF SAMPLE
WITH RESISTANCE GREATER THAN 20 MEGAOHMS,
R - .039 OHMS

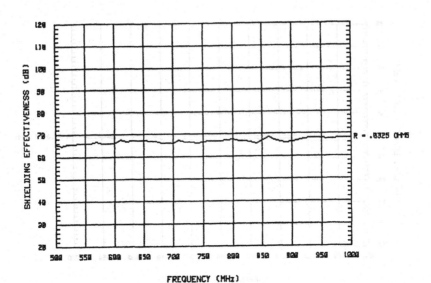

SHIELDING EFFECTIVENESS VERSUS FREQUENCY OF SAMPLE
WITH RESISTANCE GREATER THAN 20 MEGAOHMS,
R .0325 OHMS

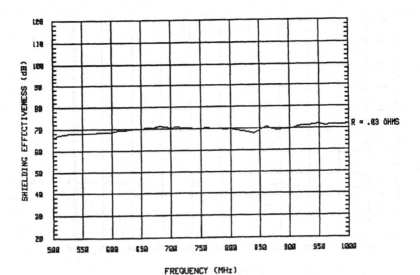

SHIELDING EFFECTIVENESS VERSUS FREQUENCY OF SAMPLE
WITH RESISTANCE GREATER THAN 20 MEGAOHMS,
R − .03

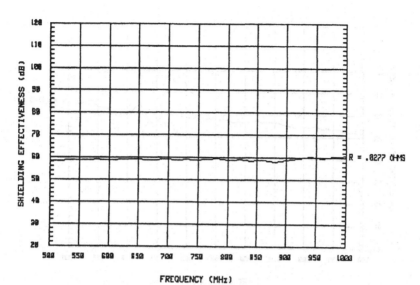

SHIELDING EFFECTIVENESS VERSUS FREQUENCY OF SAMPLE
WITH RESISTANCE GREATER THAN 20 MEGAOHMS,
R = .0277 OHMS

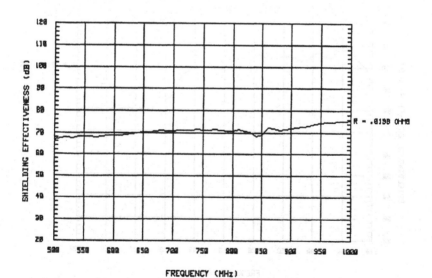

SHIELDING EFFECTIVENESS VERSUS FREQUENCY OF SAMPLE
WITH RESISTANCE GREATER THAN 20 MEGAOHMS,
R = .0198 OHMS

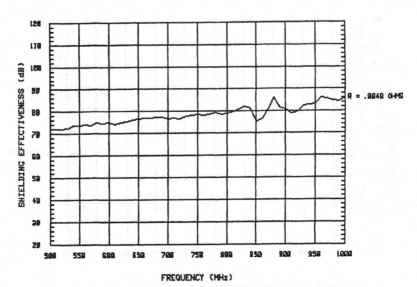

SHIELDING EFFECTIVENESS VERSUS FREQUENCY OF SAMPLE
WITH RESISTANCE GREATER THAN 20 MEGAOHMS,
R = .0048 OHMS

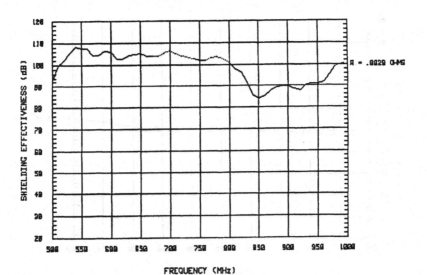

SHIELDING EFFECTIVENESS VERSUS FREQUENCY OF SAMPLE
WITH RESISTANCE GREATER THAN 20 MEGAOHMS,
R - .0029 OHMS

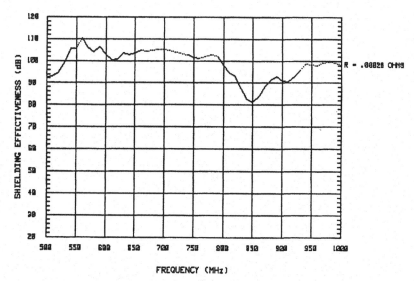

SHIELDING EFFECTIVENESS VERSUS FREQUENCY OF SAMPLE
WITH RESISTANCE GREATER THAN 20 MEGAOHMS,
R - .00028 OHMS

Appendix B
Shielding Effectiveness Versus Plate Separation Data

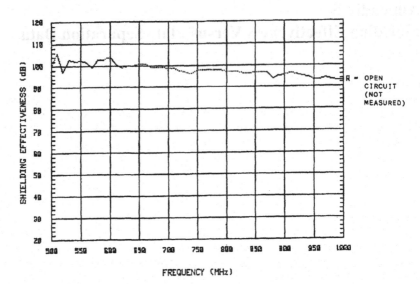

FIGURE B-1

SHIELDING EFFECTIVENESS VERSUS FREQUENCY
OF SAMPLE WITH METAL-TO-METAL CONTACT

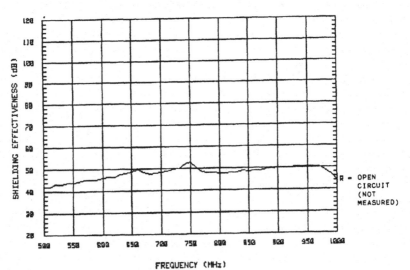

FIGURE B-2

SHIELDING EFFECTIVENESS VERSUS FREQUENCY
OF SAMPLE WITH 0.1 MM SPACING

B-2

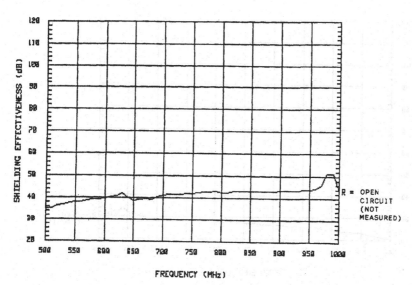

FIGURE B-3

SHIELDING EFFECTIVENESS VERSUS FREQUENCY

OF SAMPLE WITH 0.3 MM SPACING

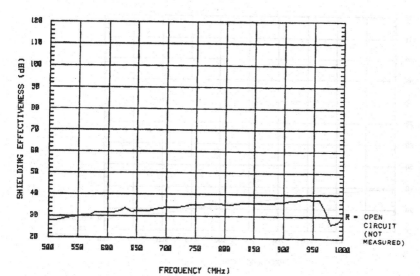

FIGURE B-4

SHIELDING EFFECTIVENESS VERSUS FREQUENCY

OF SAMPLE WITH 0.95 MM SPACING

B-3

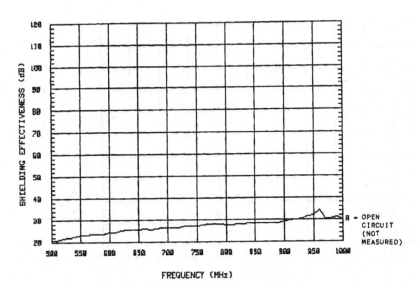

FIGURE B-5

SHIELDING EFFECTIVENESS VERSUS FREQUENCY
OF SAMPLE WITH 2.5 MM SPACING

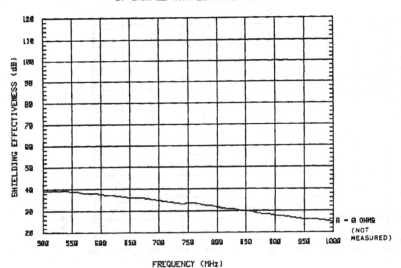

FIGURE B-6

SHIELDING EFFECTIVENESS VERSUS FREQUENCY OF SAMPLE
WITH 2.5 SPACING WITH MOUNTING SCREWS NOT INSULATED

B-4

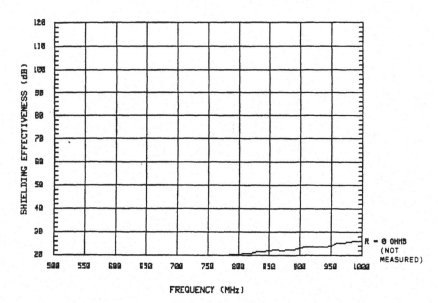

FREQUENCY (MHz)

FIGURE B-7

SHIELDING EFFECTIVENESS VERSUS FREQUENCY
OF SAMPLE WITH 2.5 SPACING AND
TWO CENTER SCREWS ON EACH SIDE REMOVED

B-5

References

1. Fontana, M. G. and Greene, N. D. (1978) *Corrosion Engineering*, McGraw-Hill Book Company, New York.
2. *National Association of Corrosion Engineers Basic Corrosion Course* (1970) National Association of Corrosion Engineers, Houston, Texas.
3. *National Association of Corrosion Engineers's Reference Handbook*, (2002) NACE International, Houston, Texas.
4. Harrington, R. F., (1961) *Time-Harmonic Electromagnetic Fields*, McGraw-Hill Book Company, New York.
5. Schelkunoff, S. A. (1943) *Electromagnetic Waves*, Van Nostrand.
6. Ott, H. W. (1976) *Noise Reduction Techniques in Electronic Systems*, John Wiley & Sons.
7. MIL-STD-285, Attenuation measurements for enclosures, electromagnetic shielding, for electronic test purposes, Method of, 25 June 1956.
8. ASTM B117, Salt spray cabinet, Method A.
9. Aerospace Material Specification 3262, February 2000, Sealing compound. silicone rubber, Two-Part, Electrically conductive and corrosion resistant (Nonchromated) for use from −55 to 260 °C (−67 to 500 °F), SAE International, Warrendale, PA, prepared by Jan W. Gooch.
10. Butler, J. (1998) EMI field notes—Design considerations for EMI gaskets. Medical Device & Diagnostic Industry Magazine. February.
11. Hudak, S., A guide to EMI shielding gasket technology, http://www.ce-mag.com/ARG/Hudak.html.

Index

A

absorption loss, 18–20
activation polarization, 9
active-passive
  metal, 8
  transition, 8–9
aerospace material specification,
  102
AFSC design handbook modification
  1–4, 91–97
alloy
  heterogeneous, 10–11
  homogenous, 10–11
aluminum
  faying surfaces, 57
  oxidation, 7
  oxide film, 5
  test joint, 35, 38–39
aluminum surfaces
  annodized conversion type, 99
  chemical conversion type, 99
aluminum-filled polyurethane RTV
  sealant, 38, 43, 45–46
anodic reaction, 7
Atlas Salt Spray (Fog) Chamber, 42
attenuation constant, 20

B

BeCu wire mesh, 104
Beryllium copper, 104
bond
  aperture dimensions, 2
  impedance of, 2
  performance, 2
  resistance, 2

C

cathodic diffusion control, 9
chloride ions migration, 13
chomerics 4375-27-4, 46–47
chromate conversion coating, 5
conductive sealant, 24, 26, 32, 35–36,
  39, 46
  in aircraft, 99
  joint, 48
conductive
  fabric-clad foam gaskets, 105
  gel gaskets, 106
copper-clad steel, 104
corrosion
  between metal surfaces, 2
  conditions for, 6–8
  definition of, 5
  electrochemical nature of, 6

prevention of, 99
rate, 8–9, 12
resistance of, 5
types of, 11
corrosion prevention
importance, 1–3
materials, 99
crevice corrosion, 11–13

D

DC resistance, 32
measurements of, 28
relationship with shielding effectiveness, 25, 30
stainless steel test joint for, 25–26
test results, 50
diffusion control, 9
double kelvin bridge, 28

E

E-3A aircraft field test
evaluation, 57–73
results, 72
E-3A aircraft
testing, 59, 63
testing results, 72
test sites, 63
effective electrical bonding, 2
elastomeric gaskets conductively
loaded, 105
electric field shielding effectiveness,
17
electrical bonding, 2
electrochemical aspect, 6
electrochemical reactions, 6–7
passivity, 8
polarization, 7–8

electrochemical corrosion of zinc, 7
electromagnetic
compatibility, 91
energy, 88
environment, 1
field, 17–18
interference, 1
pulse, 1
electromagnetic shield effectiveness,
17
electromagnetic shielding
examples of, 1–2
importance of, 1–3
process, 18
role of conductive sealants in, 24
theory, 18
electroneutrality, 14
EMI gasket
applications, 103
selection criteria, 103
EMI *see* electromagnetic interference
EMI/EMP hardness, 27, 29, 31–32
EMP *see* electromagnetic pulse
EMP/EMI requirements, 57
environmental aspect, 6, 8–10
effect of oxygen, 8
effect of oxidizers, 8
effects of velocity, 9
temperature, 9
effects of corrosion concentration,
9
galvanic coupling, 10
effect of microorganisms, 10
erosion corrosion, 9, 11, 14
ESE *see* electric field shielding
effectiveness
ethylene propylene diene monomer,
105

F

Field test
  description, 63
  method, 63
filiform corrosion, 13
fluorosilicone, 105

G

galvanic attack, 11–12

H

high intensity radiated fields, 1
high power microwave, 1
HIRF requirements, 1
HIRF *see* High intensity radiated
  fields
hollow-core gaskets, 104
HPM *see* High power micro-wave

I

induced output voltage, 48
inhibitive sealant shielding effectiveness, 101
inter-granular corrosion, 11, 14

K

Kelvin bridge milliohmmeter, 42, 47

M

magnetic field shielding effectiveness, 17
metal chloride concentration, 13
metal gaskets
  formed, 103
  stamped, 103

metallurgical aspect, 10–11
MIL-B-50878 Class R bonding
  requirement
  assessment of validity, 75–102
MIL-B-5087B modification, 80–82
MIL-HDBK-253 modification, 88–91
military specification, 80
military standard, 82, 85, 87
2.5-milliohm bonding requirements,
  76–77
MIL-STD-1541 modification, 85–86
MIL-STD-1542 modification, 86–87
MIL-STD-188-124A modifications,
  82–85
MIL-STD-454K modification, 87
Monel, 104
MSE *see* Magnetic field shielding
  effectiveness
multiple reflection correction term,
  18, 21

N

NATO STANAG No. 3659
  modification, 97–98
NATO Standardization Agreement,
  97
near field sources effect, 21–22
network analyzer, 48
nickel-filled polysulfide sea-lant, 38,
  43, 46

O

opening resealable joints, 101
optimum conductive sealant materials
  evaluation of, 38
  identification of, 35
oxidation reaction, 7
oxidizer additions effect, 8
oxygen depletion, 12–13

P

passification, 8
permeability of free space, 20
permittivity of free space, 20
pit depth, 14
pitting, 11, 13–14
plane wave shielding effectiveness, 18
polarization
  activation, 7
  concentration, 7
  concept, 7
practical electromagnetic shielding, 22–23
PRC 1764 A-2, 46–47
PRC 1764 Class B, 46–47
propagation constant
  of field, 19
  of wave inside shield, 20
propagation medium characteristic impedance, 20

R

radian frequency, 20
radio frequency, 1
reference channel of network analyzer, 48
reflection loss, 18, 20–21
relationship between
  DC Resistance, 46–53
  shielding effectiveness, 46–53
  transfer impedance, 46–53
resealable joints
  conductive sealants, 100
  in salt spray chamber, 100
RF bond impedance, 45
RF *see* radio frequency

S

salt fog environment, 36
salt spray environment per ASTM B117, 38
salt spray exposure tests, 42, 43–45
Schelkunoff's formulation, 18
SE *see* shielding effectiveness
sealant test samples evaluation, 46
selective leaching, 11, 14
shield conductivity, 20
shield effectiveness theoretical, 22
shield material intrinsic impedance, 20–21
shield permeability, 20
shielding barrier, 23
shielding effectiveness
  effect of plate separation on, 33
  measurements, 29
  test setup, 29–30
  test results, 33, 50
  value, 31
shielding effectiveness, 17, 23
silver layer, 36
skin depth, 20
stainless steel test joint, 26
stress corrosion cracking, 14
stress corrosion, 11, 14
surface current density, 23, 48

T

test sites location, 63
transfer impedance
  of joint, 23
  measurements, 48
  test results, 52
transmission coefficient, 21
transmission line analogy, 18

U

uniform attack, 11

V

voltage drop across joint, 23

W

wave impedance, 22
weathering results, 42
wire mesh gaskets, 104